Federal Highway Administration

Integrating Climate Change into the Transportation Planning Process

Final Report

July 2008

Prepared for
Diane Turchetta

Prepared by
ICF International

Table of Contents

1. Introduction 1
   1.1. Purpose 1
   1.2. Methodology 1
2. Opportunities for Linking Transportation Planning and Climate Change 3
   2.1. Review of the Long-Range Transportation Planning Process 3
   2.2. Relationship of Federal Planning Statutes and Regulations 6
3. Overview of Current Practice 11
   3.1. Summary of Climate Change Integration Activities 11
   3.2. Influence of State Climate Change Regulatory Action 12
4. Inclusion of Climate Change in Existing Transportation Plans 14
   4.1. Statewide Plans 16
   4.2. Regional Plans 18
   4.3. Barriers and Needs 20
5. Quantification of GHG Impacts in Transportation Plans 21
   5.1. Transportation Plan GHG Impacts – Current Practice 22
   5.2. Transportation GHG Inventories – Current Practice 23
   5.3. GHG Impacts from Construction – Current Practice 24
   5.4. Barriers and Needs 25
6. GHG Mitigation Strategies in Transportation Planning 27
   6.1. Land Use Integration 27
   6.2. Pricing and Alternative Modes 28
   6.3. GHG Benefits of Existing Strategies 29
   6.4. Barriers and Needs 29
7. Climate Change Adaptation in Transportation Planning 31
8. Institutional Relationships 33
   8.1. Current Practices 33
   8.2. Barriers and Needs 34
9. Conclusion 35
Appendix A: List of Interviewees 37
Appendix B: Opportunities for Integrating Climate Change Considerations into Federal Transportation Planning Regulations 39

List of Figures

Figure 1: Opportunities to Integrate Climate Change in Long Range Transportation Planning 5

List of Tables

Table 1: Applicability of Federal Planning Factors to Climate Change (23 CFR 450.206(a) and 450.306(a)) 8
Table 2: State DOT Integration of Climate Change in Long-Range Planning Documents 14

Table 3: MPO Integration of Climate Change in Long-Range Planning Documents 15
Table 4: State and MPO Quantification of GHG Emissions from Transportation Plans 21
Table 5: Comparison of MTC Plan Alternatives: 2035 CO2 Emissions (Thousand Tons/Day) ... 28
Table 6: MPOs and DOTs Involved in a Multi-agency Climate Change Initiative or Plan 33

1. Introduction

1.1. Purpose

There is general scientific consensus that the earth is experiencing a long-term warming trend and that human-induced increases in atmospheric greenhouse gases (GHGs) are the predominant cause. The combustion of fossil fuels is by far the biggest source of GHG emissions. In the United States, transportation is the largest source of GHG emissions, after electricity generation. Within the transportation sector, cars and trucks account for a majority of emissions.

Opportunities to reduce GHG emissions from transportation include switching to alternative fuels, using more fuel efficient vehicles, and reducing the total number of miles driven. Each of these options requires a mixture of public and private sector involvement. Transportation planning activities, which influence how transportation systems are built and operated, can contribute to these strategies.

In addition to contributing to climate change, transportation will likely also be affected by climate change. Transportation infrastructure is vulnerable to predicted changes in sea levels and increases in severe weather and extreme high temperatures. Long-term transportation planning will need to respond to these threats.

The objective of this study is to advance the practice and application of transportation planning among state, regional, and local transportation planning agencies to successfully meet growing concerns about the relationship between transportation and climate change. This report explores the possibilities for integrating climate change considerations into long range transportation planning at state DOTs and MPOs. The report reviews the experience of a number of DOTs and MPOs that are already incorporating climate change into their transportation planning processes and identifies their successes as well as challenges faced by these agencies.

1.2. Methodology

Our research for this report was conducted in four stages. First, we reviewed typical long-range transportation planning practices. We examined the inputs and outputs of each major step in transportation planning for relevance to climate change planning. We consider broadly how transportation planning could be enhanced to take account of GHG emissions and the risk posed to transportation systems by climate change.

Second, we reviewed federal regulations and statutes that govern transportation planning. While current regulations do not mention climate change or GHG emissions, parts of regulations can be interpreted as relevant to climate change. We point out these interpretations and general opportunities to link federal regulation of transportation planning to climate change.

Third, we reviewed a sample of current planning documents from state DOTs and MPOs for incorporation of climate change. We initially reviewed documents from 12 DOTs and 18 MPOs nationwide. We included a mix of small and large organizations from all regions of the country. We analyzed both long-range transportation plans (LRTPs) and related documents for integration of climate change.

Finally, we conducted interviews with four DOTs and eight MPOs that are working to incorporate climate change into long range transportation planning. We probed for successes, barriers and solutions, and common approaches. We identified trends across agencies' experiences with planning for climate change.

Note that opinions stated in this memo represent the viewpoints of individual staff members at DOTs and MPOs and are not necessarily official agency positions.

The remainder of this report presents our research findings. Section 2 contains our review of planning practices and federal regulations. Section 3 presents a summary of current practices among state DOTs and MPOs. Section 4 discusses the integration of climate change into the text of LRTPs. Each of Sections 5-8 focuses on a specific trend or topic area in planning for climate change. We describe current practices, barriers, and needs, and provide supporting examples from specific DOTs and MPOs. Section 9 discusses issues for future research.

2. Opportunities for Linking Transportation Planning and Climate Change

State DOTs and MPOs nationwide use common methods in transportation planning. Most long range transportation planning efforts follow a sequence of several steps to arrive at a plan. At each step certain information is incorporated and decisions are made. While these typical practices may not be explicitly mentioned in federal planning statues and regulations, they are conventional applications that are common across regional and state boundaries. Within these conventions, there are potential points to integrate considerations of climate change. Although there is no federal mandate to consider climate change in transportation planning, state DOTs and MPOs can go beyond existing statutes to integrate climate change at these specific points.

At the same time, federal planning legislation plays an important role in transportation planning processes. DOTs and MPOs are required to meet certain minimum standards in their LRTPs. Federal regulation establishes planning timelines, factors to consider, and parties to consult. While future legislation may explicitly mention climate change and GHG emissions, existing legislation is already relevant to climate change. Both legislators and planners should be aware of how federal regulations can enable consideration of climate change in long range transportation planning.

2.1. Review of the Long-Range Transportation Planning Process

Transportation plans can consider both mitigation of and adaptation to climate change. Mitigation of climate change means reducing the major cause of climate change: GHG emissions released by human activities. Adaptation to climate change means minimizing the potential impacts on the transportation system from climatic changes such as rising average temperatures, increased intensity of storms, rising sea levels, and increases in overall climatic variability.

Opportunities for practitioners to address climate change exist within several key elements of transportation plans. Each of these elements can incorporate climate change directly, by explicitly addressing climate change, and indirectly, by addressing elements of transportation that are linked to climate change. Components within both statewide and metropolitan transportation plans that can include climate change are:

- **Vision and goals** – GHG emission reductions and related climate change mitigation could easily become a stand-alone goal in transportation plans. Planners can also incorporate climate change indirectly by emphasizing linkages between climate change and existing plan goals. Many transportation plans already include goals that address environmental issues that might relate to energy and climate change.

- **Trends and challenges** – Rising GHG emissions from transportation and looming threats to the system from the impacts of climate change are important long term trends. Issues to be considered include VMT growth, congestion, changing development and land use patterns, sea level rise, accelerated aging of infrastructure from climate change, and rapidly changing fuel and vehicle technologies.

- **Strategies and improvement projects** – Strategies and improvement projects that target climate change are essential to the long term performance of the transportation system. Most demand management and system management strategies reduce GHG emissions. Other types of strategies can reduce the risk from flooding associated with climate change. Plans can include new strategies targeted at GHG emission reductions and adaptation to climate change, as well as link existing strategies to climate change.

- **Performance measures** – Performance measures can assess whether or not objectives related to climate change are met. Performance measures can be unique to climate change and energy efficiency goals (for example, GHG emissions per capita, petroleum use per VMT, percent of alternative fuel vehicles) or relate to traditional transportation planning goals such as congestion or air quality (for example, transit mode share, average vehicle occupancy). Performance measures can be used to evaluate the existing system, compare and select alternatives, and measure the progress of the plan throughout its implementation. In addition, performance measures can assist in prioritizing projects for programming in the TIP.

In addition to the basic components of planning *documents*, transportation agencies can also incorporate climate change through their long term planning *processes*. The stakeholders involved, the information considered, and the decision-making structure at each step of transportation planning influence the eventual outcome of the plan. Opportunities to incorporate climate change throughout the transportation planning process include:

- **Coordinate:** Many of the agencies and stakeholders that DOTs and MPOs already work with as interested parties (23 CFR 450.316(a)) may have particular interests in climate change or environmental issues. Potential stakeholders include environmental agencies and interest groups, local air quality or natural resource agencies, freight carriers (including trucking, rail, marine, and airline companies), operating agencies (including toll-way authorities), local government representatives, and safety, security, and emergency response teams. These stakeholders can bring new expertise and resources to the planning process. Particularly, stakeholders involved in climate action planning at the state or metropolitan level can help coordinate transportation planning with those efforts.

- **Integrate land use:** The promotion of compact and transit-oriented development patterns is potentially one of the most effective strategies to reduce GHG emissions from transportation in the long-term, but it also requires a great degree of collaboration among agencies and among plans. While transportation planning has long considered future land use patterns in the development of travel demand forecasts, there has been less success in ensuring that transportation investment decisions support a regional vision for growth. Transportation planning can consider cross-linkages with land use plans and involve agencies with jurisdiction over land use plans.

- **Link funding:** Beyond the long range plan, the transportation improvement program (TIP) details what near-term projects are going to be built and when, based on funding cycles. Transportation agencies can use performance measures related to climate change to prioritize projects for funding within the transportation improvement program. The performance measures should be based upon the strategies and objectives in the long-range plan that relate to climate change. Adaptation as well as mitigation can be considered in plan development and programming. In long-range planning, performance measures related to adaptation may include the location and types of system networks; in programming, adaptation may focus more on determining project costs from location and design constraints.

The figure below illustrates the basic steps in the long range transportation planning process of state DOTs and MPOs in relation to the above opportunities to integrate climate change. The steps are organized to reflect the typical metropolitan transportation planning process. Most MPOs will go through each of these steps in drafting their transportation plans, although the order of steps may vary slightly. In addition, state DOTs use many of the same steps in drafting statewide transportation plans. The steps can also apply to the drafting of corridor plans, an increasingly important function of MPOs. The paragraphs following the figure explain the steps and their relation to climate change in more detail.

Figure 1: Opportunities to Integrate Climate Change in Long Range Transportation Planning

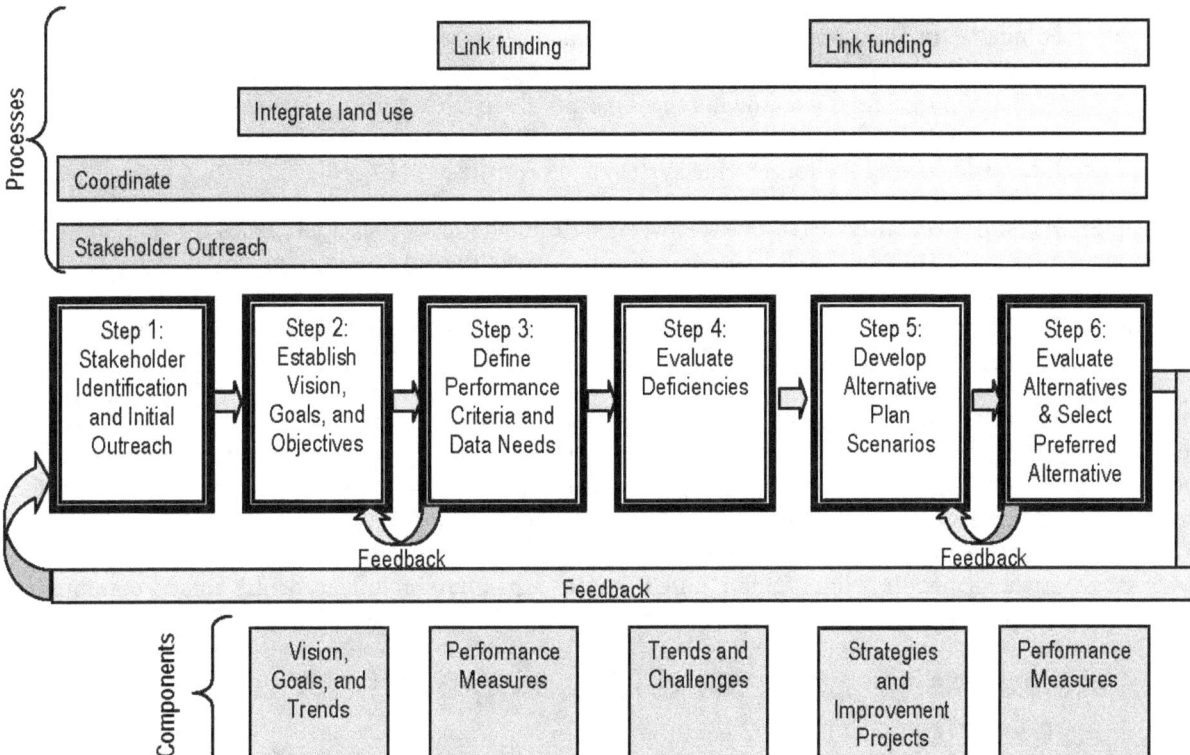

At **Step 1: Stakeholder Identification and Initial Outreach**, transportation agencies typically identify partners and stakeholders to participate in the process. They also establish and publicize a framework for the planning process. Although outreach and coordination continues throughout the planning process, this is a key point for its initiation. Transportation agencies can actively engage environmental and state and local government agencies and other organizations involved in climate action planning. Private industries with a large impact on transportation GHG emissions, such as those with large vehicle fleets, can also be contacted.

At **Step 2: Establish Vision, Goals, and Objectives**, agencies establish a vision and related goals and objectives for the transportation system. The vision reflects an overall desired end state for the system, while goals and objectives address individual aspects of the system such as mobility, safety, preservation, and environmental performance. The vision can emphasize mitigating the system's impact on climate change and consider preservation of the system in the face of shifts in climate. Goals can address specific linkages between transportation and climate change such as VMT per capita, total energy use, and total GHG emissions. Connections between transportation and land use can be formally considered beginning with this step.

In **Step 3: Define Performance Criteria and Data Needs**, agencies develop criteria that will measure progress towards stated objectives. They also identify data that will be needed to measure performance against the criteria. Agencies can include performance measures related to GHG emissions and adaptation to climate change. The selection of appropriate performance measures at this step can help prioritize funding for projects that mitigate or adapt to climate change at later stages of the process.

In **Step 4: Evaluate Deficiencies,** agencies characterize the existing system relative to performance criteria, gather input from stakeholders and the public on priority deficiencies, and forecast future deficiencies. At this stage agencies can incorporate climate change into the assessment of trends and challenges. Specifically, agencies can recognize and assess the threat posed by GHG emissions and climate change and forecast vulnerability of the transportation system to climate change. This step might involve screening existing transportation infrastructure and proposed projects to determine their potential vulnerability to the impacts of climate change. Once this screening step is complete, it may be necessary to conduct detailed vulnerability assessments that incorporate localized projections of climate change impacts and the associated effects on the system. Additionally, adaptation also may be an important criterion both for determining the form of the system and prioritization of projects.

At **Step 5: Develop Alternative Plan Scenarios,** agencies develop various approaches for achieving the stated objectives and distill several diverse, manageable alternatives. Agencies typically identify fiscal constraints and opportunities at this stage as well. This step can identify alternatives that facilitate mitigation of GHG emissions or adaptation to climate change. Specific strategies and improvement projects can be included in the alternatives developed.

In **Step 6: Evaluate Alternatives and Select Preferred Alternative,** agencies apply performance criteria to alternative scenarios and select the alternative that best meets community goals. This stage includes the preparation of a tiered project list based on the selected scenario. At this step, agencies can apply climate change related performance criteria to the alternatives developed. Agencies can examine the impacts of proposed mitigation and adaptation strategies to ensure that the selected alternative appropriately addresses climate change. It is important to note that decisions often include tradeoffs among community goals. For example, maximizing highway mobility may increase GHG emissions.

2.2. Relationship of Federal Planning Statutes and Regulations

Several federal statutes and regulations govern the transportation planning process. The text of these documents provides some opportunities to link climate change considerations with the planning process. Specifically, federal planning regulations include:

- Title 23 USC 134-135 (FHWA)
- 23 CFR Parts 450 and 500 (FHWA)
- Title 49 USC 5303 (FTA)
- 49 CFR Part 613 (FTA)

These documents set out the requirements for undertaking transportation planning, and include statements regarding the scope of planning processes, required procedures, and required content for metropolitan and statewide transportation planning under both FHWA and FTA. While there are no specific requirements to directly address climate change, recent revisions to legislation have further incorporated energy and environmental considerations. These revisions offer greater opportunities for MPOs and state DOTs to integrate climate change considerations within their planning processes. For example, 23 USC 143(a) states that it is in the national interest to:

> *... encourage and promote the safe and efficient management, operation, and development of surface transportation systems that will serve the mobility needs of people and freight and foster economic growth and development within and between States and urbanized areas, while*

minimizing transportation-related fuel consumption and air pollution *through metropolitan and statewide transportation planning processes...*

The goal of minimizing fuel consumption and air pollution can be interpreted as a direct link to climate change and justification for metropolitan transportation planning to consider climate change mitigation strategies. Section 23 USC 135(a) mandates similar consideration of fuel-consumption and air pollution in statewide planning. Additionally, requirements for the examination of the effects of transportation decisions on the environment and energy consumption are reiterated throughout the planning legislation. Energy and environment are one of the eight required planning factors.

The planning regulations also include a number of requirements that generally align with climate change mitigation and adaptation. For example, provisions that relate to efficient management and operation of the transportation system, coordination with land use plans, and congestion mitigation can all related to reducing GHG emissions. For adaptation, the requirements for infrastructure preservation and maintenance, as well as corridor preservation and connectivity of the system, can provide direct avenues for consideration of adaptation strategies in planning.

Types of Linkage Opportunities

There are four general types of climate change linkage opportunities in planning statutes and regulations:

1. **Requirements to address energy and environmental concerns** – These sections provide a link to GHG mitigation, since GHG emissions from transportation are largely correlated with energy consumption and impacts of climate change are important considerations in environmental protection. (23 CFR 450 Subparts 200, 206, 214, and 306)

2. **Requirements to ensure an integrated transportation system, preserve the projected and existing system, and ensure the safety and security of the system for users** – These sections could be interpreted as requiring or encouraging adaptation strategies, since MPOs and state DOTs will need to consider the implications of climate change (such as sea level rise and accelerated aging from temperature swings) on their infrastructure to ensure effective connectivity is preserved. (23 CFR 450 Subparts 206, 214, and 306; 49 CFR 613 Subparts 100 and 200)

3. **Transportation demand management and transportation system management strategies** – Many sections of the federal requirements contain language that encourages these strategies. Both can be considered climate change mitigation strategies, if they reduce congestion or reduce vehicle travel. Similarly, transit enhancements and emphasis on non-motorized (pedestrian and bicycle) transportation can potentially serve as climate change mitigation strategies. (23 CFR 450 Subparts 200 and 320)

4. **Consultation requirements** – These requirements could be interpreted as requiring that transportation planning processes consider climate action planning activities going on at the state or regional level, local government plans or policies that may consider climate change, and the work of environmental agencies as it relates to climate change and GHG emissions. (23 CFR Subpart 208 and 214)

Federal Planning Factors

Federal statutes require MPOs and state DOTs to consider eight factors in their planning functions. These factors serve as de facto goals for transportation planning and include topic areas of economic vitality, safety, security, mobility, environment, connectivity, efficiency, and preservation. While climate change is most directly related to the environment and energy conservation planning factor (factor 5), it is related to each of the eight factors, if indirectly, as illustrated in Table 1.

Table 1: Applicability of Federal Planning Factors to Climate Change (23 CFR 450.206(a) and 450.306(a))

Planning Factor	Applicability of Climate Change Considerations
(1) support the economic vitality of the [United States, the States, nonmetropolitan areas, and] metropolitan area[s], especially by enabling global competitiveness, productivity, and efficiency;	In addition to a physical threat, climate change also poses an economic threat. Climatic changes can damage natural environmental assets as well as manmade assets. Weather-related natural disasters can cause damage worth billions of dollars. These losses have a direct toll on local, regional, and national economies. At the same time, the development of new technology to reduce and prepare for climate change offers economic development opportunities. New transportation technologies can generate new economic activity as they are developed and exported.
(2) increase the safety of the transportation system for motorized and nonmotorized users;	A safe transportation system protects users from hazards, including hazards resulting from climate-related stresses on the system. Transportation agencies need to protect the system from potential floods and perform routine maintenance and replacement on infrastructure components affected by extreme temperatures and storms. Other safety enhancements can actually reduce GHG emissions. Enhancements that reduce the risk of crashes and smooth traffic flow reduce GHG emissions from congestion. In some cases, slowing vehicle travel speeds can contribute to improved fuel efficiency and improved safety.
(3) increase the security of the transportation system for motorized and nonmotorized users;	A secure transportation system ensures the protection of critical infrastructure and exposes users to less risk. Infrastructure protection is going to require assessing risk from climate-related stresses on the system. Transportation agencies need to consider security as part of a broader consideration that incorporates planning for natural disasters, emergency response and preparedness and infrastructure preservation.
(4) increase the accessibility and mobility of people and freight;	While accessibility and mobility have often been interpreted as synonymous with more travel by car and truck, these goals can also be achieved with reduced vehicle travel. Multimodal transportation systems can be coordinated with land use patterns such that people and goods need to travel shorter distances and make fewer trips by car and truck. In fact, travel by private car is inherently inaccessible for many low-income, elderly, and young people. The systematic provision of other options both improves mobility for these populations and helps to reduce GHG emissions.

Planning Factor	Applicability of Climate Change Considerations
(5) protect and enhance the environment, promote energy conservation, improve the quality of life, and promote consistency between transportation improvements and State and local planned growth and economic development patterns	Mitigating climate change is essential in order to protect the environment from long term shifts in weather patterns. Reducing GHG emissions is virtually equivalent to conserving energy, since most GHG emissions come from the burning of fossil fuels. One of the chief ways that transportation agencies can reduce GHG emissions is to reduce the total amount of on-road travel. When transportation improvements are coordinated with planned growth patterns, the need to travel (and especially the need to travel by car) can be reduced.
(6) enhance the integration and connectivity of the transportation system, across and between modes [throughout the State], for people and freight;	One of the chief ways that transportation agencies can reduce GHG emissions is to reduce the total amount of on-road travel. Shifting passenger trips from cars to public transportation, biking, and walking, and freight trips from trucks to rail (and possibly ships) can help to reduce on-road travel. To the extent that agencies can provide more modal choices and improve the ease of transfers between modes, passengers and shippers are more likely to choose an alternative mode for at least part of each trip.
(7) promote efficient system management and operation	The energy efficiency of the transportation system depends in part on the efficient operation of the system. Travel times can be improved and congestion reduced in many cases through better incident management, real-time information distribution, and traffic flow engineering. Reduced congestion translates to improved fuel efficiency and reduced GHG emissions.
(8) emphasize the preservation of the existing transportation system	The transportation system, like other assets of our built environment, is threatened by climate change. Adaptive responses to increased heat, rising sea levels, and higher incidences of flooding must be considered in order to preserve the system.

Other Federal Regulations

In addition to federal requirements that directly address the transportation planning process, air quality and climate change issues are becoming more prevalent in the transportation planning process through other avenues. For example, a new requirement in the Energy Bill, Section 55601(f), encourages state and local governments to include short sea shipping and other marine transportation solutions in their transportation planning process as a means to reduce transportation-related energy consumption:

§55601. Short sea transportation program

(f) *Multistate, State and Regional Transportation Planning* - The Secretary, in consultation with Federal entities and State and local governments, shall develop strategies to encourage the use of short sea transportation for transportation of passengers and cargo. The Secretary shall—

(1) assess the extent to which States and local governments include short sea transportation and other marine transportation solutions in their transportation planning;

(2) encourage State departments of transportation to develop strategies, where appropriate, to incorporate short sea transportation, ferries, and other marine transportation solutions for regional and interstate transport of freight and passengers in their transportation planning; and

(3) encourage groups of States and multi-State transportation entities to determine how short sea transportation can address congestion, bottlenecks, and other interstate transportation challenges.

Similarly, SAFETEA-LU Section 6001 requires consultation and environmental mitigation strategies to be incorporated early into the transportation decision making process. This requirement presents an opportunity for considering climate change mitigation strategies. Mitigation strategies could include, for example, short sea shipping strategies, thereby also fulfilling the Section 55601(f) requirements.

Specific Linkage Opportunities

Appendix B presents specific opportunities to link existing federal transportation planning requirements to climate change mitigation and adaptation. The table in Appendix B lists the relevant sections of legislation, the aspect of planning that the section addresses, excerpted relevant language from the legislation, and the potential ties to climate change considerations. Since this table breaks down the legislation comprehensively section by section, and the language within the legislation is often repetitive between sections, the table intentionally reiterates similar opportunities for integration of climate change considerations by both MPOs and state DOTs. Within the table, we identify opportunities for direct and indirect linkages to existing language and opportunities to add new language to the legislation to create linkages.

3. Overview of Current Practice

A number of state DOTs and MPOs are already integrating climate change into planning documents and processes through the opportunities described above, and through other means. Practices are evolving rapidly. The experiences of these transportation agencies can inform other DOTs and MPOs as well as contribute to broader discussions on how best to integrate climate change into long range transportation planning.

3.1. Summary of Climate Change Integration Activities

Within long-range transportation plans, climate change appears in a number of places. The simplest way for a plan to incorporate climate change is to acknowledge its connection to transportation, often in a section on trends and challenges. Many plans also now include mitigation of GHG emissions in a hierarchical policy structure (Vision – Goals – Policies – Strategies). The position of climate change within the structure varies from plan to plan. Some plans include an individual goal related to climate change. Others include policies or strategies related to climate change mitigation under one or several goals. Finally, a few plans incorporate performance measures based on GHG emissions.

The integration of climate change into long-range transportation plans is a new development. Plans adopted within the last year or two are much more likely to incorporate climate change than plans adopted a few years ago. Many DOTs and MPOs are working on plans or concepts related to climate change that have not yet been fully incorporated into LRTPs.

Quantification of GHG emissions will likely be a key component of transportation planning in the future. In order to effectively reduce GHG emissions, planners need to know current and future emissions levels and the potential impacts of various policies and strategies on emissions. While only New York State currently requires quantification of GHG emissions from plans or transportation improvement programs (TIPs), several other states are considering such a requirement. Likewise, only a handful of MPOs have conducted analyses of GHG emissions, but more are considering or planning to conduct such analyses. Agencies face a number of questions about appropriate tools, methodologies, and data.

As expected, larger MPOs are typically doing more than smaller MPOs with regard to climate change. In addition to the obvious differences in staff sizes and resources, the variation in autonomy and regional planning powers that MPOs have also plays a role. Some larger MPOs serve as independent regional planning agencies, with influence (though not necessarily jurisdiction) over land use, economic development, natural resources, and community development. These functions allow for integrated planning, which can be much more effective at reducing GHG emissions from transportation than the traditional transportation planning process alone. The needs and concerns of small MPOs in relation to climate change are substantially different than those of large MPOs, and merit special attention.

A few GHG mitigation strategies are emerging as particularly popular among transportation agencies. Many agencies cite integrated transportation and land use planning as the most promising long term strategy for reducing transportation GHG emissions. In the shorter term, pricing strategies and promotion of alternative modes are also priorities for many agencies. Meanwhile, LRTPs often do not explicitly recognize the effect that existing policies and strategies can have to reduce transportation GHG emissions. For example, transportation demand management policies to reduce congestion and air pollution can also reduce GHG emissions.

Mitigating responses to climate change are much better developed than adaptive responses. Some agencies are beginning to develop very specific strategies to analyze and reduce GHG emissions. At this point, adaptive responses in plans are typically just acknowledgements of risk or suggestions for research.

At least one transportation agency, the San Francisco Bay Area's Metropolitan Transportation Commission, is conducting a research study on the topic.

Many DOTs and MPOs are working across institutional boundaries on climate change issues. DOTs and MPOs are both leading and participating in multi-agency climate change groups focused on developing action plans, tools, and guidance documents for climate change mitigation and adaptation. Some DOTs and MPOs are also addressing climate change through existing interagency groups. Climate change issues span boundaries of geography and jurisdiction. Many agencies recognize that multi-agency action has the greatest potential to incorporate climate change into transportation planning.

By the same token, some transportation agencies are deferring individual action on climate change in expectation of direction from multi-agency groups or from higher levels of government. There is considerable uncertainty about how climate change regulation at both the federal and state levels will impact transportation planning requirements. Many agencies are eager to receive appropriate guidance before taking steps on their own.

3.2. Influence of State Climate Change Regulatory Action

Many of the DOTs and MPOs that have been most proactive in the area of climate change are in fact driven by state legislation on GHG emissions. California in particular is developing a strong administrative and regulatory framework to reduce GHG emissions in response to AB32. The California Department of Transportation (Caltrans) and the state's large MPOs are moving rapidly to adapt to expected future mandates.

In California, Assembly Bill (AB) 32 (the Global Warming Solutions Act of 2006) drives much of the activity around climate change mitigation. The law mandates that California reduce GHG emissions to 1990 levels by 2020. The transportation sector, as a large contributor to GHG emissions, is expected to produce significant reductions. Proposed legislation (SB 375) would require the California Air Resources Board (CARB) to set emission reduction targets for the transportation sector and allocate those reduction targets to MPOs. If enacted, this legislation would directly affect the long-range planning functions of MPOs in the state.

While Congress was clear that it was not extending NEPA requirements to transportation plans and programs (23 CFR Appendix A to Part 450), recent legal and legislative action at the local level suggests that transportation agencies will also have to consider GHG emissions under the California Environmental Quality Act (CEQA). In 2007, the state Attorney General began warning MPOs that they could be required to analyze the GHG impacts of their regional transportation plans under CEQA. Even though no official guidance exists on how such analyses should be conducted, or how significance of impacts should be determined, some MPOs have begun exploring GHG impacts in their CEQA documents.

In addition, the California Transportation Commission (CTC) recently proposed revisions to California's RTP Guidelines that incorporate climate change into the regional transportation planning process. The new guidelines suggest specific policies, strategies, and performance measures for regional smart growth. They suggest modeling and analysis techniques for transportation GHG emissions. They also promote technical assistance by Caltrans and the CTC to regional transportation planning agencies for GHG modeling.

Washington State is also instituting regulations that will affect transportation planning. In March 2008, Washington Governor Gregoire signed climate change framework legislation HB 2815, which includes a requirement to reduce light duty vehicle per capita VMT 18% by 2020, 30% by 2035, and 50% by 2050. Washington has formed a Climate Action Team (CAT), which includes Washington's Transportation

Secretary. The CAT is being advised by sector-specific Implementation Working Groups, charged with developing the specific implementation steps needed to achieve the reductions in the climate plan. A WSDOT representative co-chairs the Transportation Implementation Working Group. The group is currently developing recommendations for specific actions that can help achieve the VMT reduction goals laid out in HB 2815. It is likely that implementation steps will eventually result in new requirements on LRTPs in Washington.

In New York State, the State Energy Plan, adopted in 2002, is one of the first in the nation to integrate transportation planning, energy conservation, greenhouse gas mitigation, and air quality planning. One of the recommendations in the State Energy Plan is that MPOs, in conjunction with the State, assess the energy use and greenhouse gas emissions expected to result from implementation of transportation plans and programs. In response, NYSDOT has drafted methodological guidance to help MPOs fulfill this recommendation. All 13 MPOs in New York have now estimated energy use and CO_2 emissions from their LRTPs and also from their transportation improvement programs (TIPs).

4. Inclusion of Climate Change in Existing Transportation Plans

DOTs and MPOs can include climate change in the text of LRTPs to a variety of degrees. Climate change can appear in the vision, goals, policies, strategies, trends and challenges, and performance measures promulgated by LRTPs. Some plans merely recognize that climate change is an issue that relates to transportation and begin to point out the relevance of existing plans and strategies to climate change. Other plans make climate change more central to their goals and policies. Some plans reference other documents that deal with climate change, while others propose to act as drivers of climate change policy.

Ultimately, long range planning involves much more than just producing planning documents. This chapter focuses on inclusion of climate change in the text of LRTPs. Subsequent chapters will expand on how MPOs and DOTs are incorporating climate change in specific plan elements and related processes.

Tables 2 and 3 classify the integration of climate change in 12 state DOT and 18 MPO LRTPs reviewed for this study. Integration of climate change is classified by the plan components in which climate change is included. As closely related topics, energy conservation and alternative fuels are also classified as included in plans. Most plans include only mitigation of climate change, but a few also refer to adaptation to climate change, as indicated by an asterisk.

Table 2: State DOT Integration of Climate Change in Long-Range Planning Documents

DOT	Status of LRTP Reviewed	Climate Change Mitigation in:				Energy Conservation or Alternative Fuels in:			
		Trends and Challenges	Vision and Goals	Policies and Strategies	Performance Measures	Trends and Challenges	Vision and Goals	Policies and Strategies	Performance Measures
Maine	adopted Dec 2007	X	X	X		X	X	X	X
New Mexico	adopted 2005						X	X	
Arizona	adopted Sep 2004						X		
Colorado	draft Sep 2007	X							
Connecticut	adopted July 2004	X*	X*	X*		X	X	X	
Massachusetts	adopted 2006			X				X	X
Maryland	draft goals 2008								
Oregon	adopted Sep 2006	X*	X	X		X	X	X	
Washington	adopted Nov 2006	X					X		
California	adopted April 2006	X	X	X		X	X	X	X
Florida	adopted 2005		X				X		
New York	adopted 2006			X					

* Includes adaptation

Notes: Table 2 designates only those components that LRTPs explicitly link to the topic areas. Energy conservation does not include reference to energy conservation solely under the eight Federal Transportation Planning Factors. The table is not comprehensive; it reflects only those states reviewed as part of this study.

Table 3: MPO Integration of Climate Change in Long-Range Planning Documents

MPO Region	Status of LRTP Reviewed	Climate Change Mitigation in:				Energy Conservation or Alternative Fuels in:			
		Trends and Challenges	Vision and Goals	Policies and Strategies	Performance Measures	Trends and Challenges	Vision and Goals	Policies and Strategies	Performance Measures
Eugene, OR	final draft Sep 2007			X			X	X	X
Missoula, MT	adopted May 2004		X				X		
Santa Fe, NM	draft due 2009								
Albany, NY	draft August 2007			X					
Grand Rapids, MI	adopted April 2007	X				X	X		
Portland, OR	final draft Jan 2008	X*	X	X	X		X	X	
Salt Lake City	adopted May 2007						X		
Baltimore	adopted Nov 2007	X		X			X	X	
Chicago	updated June 2007		X				X		
Denver	adopted Dec 2007					X	X	X	X
Houston-Galveston	updated Oct 2007	X*					X		
Philadelphia	adopted 2005								
Sacramento	draft Nov 2007	X		X		X		X	
San Diego	adopted Nov 2007	X	X	X		X		X	X
San Francisco	draft goals 2008		X*		X				
Seattle	adopted Spring 2008**	X	X*	X	X			X	
Southern California	adopted May 2008	X			X	X	X		
Washington, DC	adopted Oct 2006								

* Includes adaptation
** Refers to Vision 2040. Vision 2040 is a regional growth, transportation, and economic strategy. It is not the official LRTP for the region. The official LRTP will implement the policies of Vision 2040 and is scheduled for adoption in 2010.
Notes: Table 3 designates only those components that LRTPs explicitly link to the topic areas. Energy conservation does not include reference to energy conservation solely under the eight Federal Transportation Planning Factors. The table is not comprehensive; it reflects only those MPOs reviewed as part of this study.

The DOTs and MPOs examined were selected for diversity of size and geography. We prioritized agencies that we knew to be active on climate change issues. Of the 12 DOTs whose LRTPs were examined, 9 mentioned climate change in some capacity. None of these included climate change in performance measures, but performance measures related to energy conservation were included in 3 of the LRTPs. Maine, Connecticut, Oregon, and California covered climate change most broadly across LRTP components.

Of the 18 MPO LRTPs examined, 13 mention climate change. Climate change appears variously in the trends and challenges, vision and goals, policies and strategies, and performance measures of MPOs'

LRTPs. Among the plans reviewed, Portland, San Diego, and Seattle have incorporated climate change most broadly into their LRTPs.

4.1. Statewide Plans

While the previous tables provide a general picture of where climate change appears in LRTPs, actual plans differ substantially in how they treat climate change. In some plans climate change has become a priority issue. In others climate change is mentioned as a peripheral concern. A few examples illustrate the differences between state LRTPs.

Maine DOT: Connecting Maine (Adopted December 2007)

An addendum and summary to Connecting Maine, the state's LRTP, includes a section titled "Greenhouse Gases and Global Warming." The section provides an overview of the Maine emissions inventory. The draft plan cites the need for long-term strategies including utilizing low-GHG fuel, implementing tailpipe emissions standards, slowing VMT growth, and increasing the availability of low-GHG travel choices (such as transit passenger rail, vanpools, walking, and biking). One of the main strategies emphasized in the plan is to shift freight movement from the highway to rail and marine modes.

The LRTP recognizes that "transportation investments must enable businesses and individuals to shorten their trip times and use more fuel-efficient modes of transportation." The plan also includes an estimate of the emissions impact of projects included in the plan. According to these estimates, CO_2 emissions will be reduced by 26 to 32 thousand metric tons by 2020, and by 40 to 48 thousand metric tons by 2030.

Connecticut DOT: Long-range Transportation Plan for the State of Connecticut, 2004-2030 (Adopted July 2004)

In its LRTP, Connecticut DOT (ConnDOT) provides a robust discussion of climate change, GHG emissions, and fuel consumption. Within the plan, ConnDOT has a specific section to address Environment, Energy Conservation, and Quality of Life, in which they discuss the growing trends of people and goods movement over longer, more dispersed distances and the resulting increase in energy consumption. They recognize that "Connecticut's greenhouse gas (GHG) emissions from non-renewable fuel consumption are contributing to global climate change." To address this issue and support the state's climate action plan, ConnDOT establishes the following actions related to climate change in its LRTP:

- "Support programs and efforts that focus on minimizing fuel consumption, black carbon emissions, and single-occupancy vehicle trips as well as address the environmental and health costs associated with non-renewable fuel emissions"

- "Continue to participate on the Governor's Steering Committee on Climate Change and support the Committee's efforts to implement recommendations for reducing harmful greenhouse gas (GHG) emissions generated by the State's transportation system"

- "Consider and address in design plans, the needs of owners of alternative fuel vehicles when constructing and renovating transportation facilities"

- "Encourage transportation research and projects that explore innovative solutions to GHG emissions using advanced technology, economically feasible options, and proven results for reducing emissions both in the short and long-terms"

- "Encourage research and projects that explore innovative solutions for responding to changing land and water patterns, including flooding and loss of coastline caused by increased frequency and severity of meteorological events which may affect components of the transportation system infrastructure"
- "Encourage efforts that focus on risk and response assessment, including prediction tools, products and strategies for potential maintenance, system planning, safety management and emergency preparedness issues arising from global climate change"
- "Encourage practices and policies that shorten delivery time and provide alternatives for goods movement through environmentally-friendly methods that reduce fuel consumption, such as coordinated intermodal transport"
- "Continue to implement energy performance standards for State transportation facilities, promote green building design on major capital projects, purchase environmentally preferable products, and use electronic media"

These detailed actions comprehensively address climate change considerations within the LRTP. They integrate a range of both mitigation and adaptation strategies.

ConnDOT is currently in the process of updating its LRTP. They agency anticipates releasing a draft for public comment in Spring 2008 and publishing the final version in Summer 2008. In updating the State's Long-Range Plan, ConnDOT will consider strategies and recommendations from state environmental reports such as the Connecticut Climate Change Action Plan 2005, the Connecticut Clean Diesel Plan (January 2006), and the Governor's Energy Plan (September 2006). Given the State's early leadership in addressing climate change through transportation planning, it is likely this new LRTP will further integrate climate change.

Oregon DOT: Oregon Transportation Plan (Adopted September 2006)

The Oregon Transportation Plan (OTP) recognizes the impact of transportation on climate change in its trends and challenges: "Transportation is causing global warming and other environmental degradation." It establishes global warming as a major challenge area for the transportation system. Rising sea levels and increased wave heights due to global warming could impact Highway 101, coastal ports, and other coastal transportation facilities in Oregon. Referencing the Oregon Strategy for Greenhouse Gas Reductions (2004), the plan also summarizes strategies that can reduce GHG emissions:

- alternative vehicle and fuel technologies
- more efficient land use patterns
- increasing use of public transportation, freight rail, bicycling, and walking

In its goals and strategies, the OTP incorporates all of the strategies mentioned above. Goal 4, Sustainability, explicitly references climate change. One of its stated objectives is to "Reduce emissions of greenhouse gases to reduce climate change." Among the strategies supporting Goal 4 is: "Encourage the development and use of technologies that reduce greenhouse gases."

Massachusetts Executive Office of Transportation: Commonwealth of Massachusetts Long-Range Transportation Plan (Adopted 2006)

Massachusetts' LRTP reiterates tasks from the state's Climate Protection Plan. The Executive Office of Transportation (EOT) dedicates a section of one chapter to the "Implementation of the New Massachusetts Climate Protection Plan." The plan outlines commitments from the Climate Protection

Plan and also details the EOT's alternative fuels program and the use of recycled materials in pavement as part of this commitment. The use of recycled materials in roadway construction can reduce energy use and thereby reduce GHG emissions.

The plan provides the following action items to meet the goal of Sustainability and Transportation – Minimizing Transportation's Impacts:

- "Implement the principles of the Massachusetts Climate Protection Plan"
- "Aggressively pursue the acquisition of alternative fuel vehicles and related infrastructure for all transportation agencies."
- "Continue to explore the feasibility of using recycled materials for pavements"

The plan also commits EOT to pursue further actions for climate change mitigation in the future.

4.2. Regional Plans

Like state DOTs, MPOs can integrate climate change in different ways into the text of their LRTPs. As federally designated agencies, MPOs are a more diverse group than are state DOTs. The jurisdictions of MPOs range in size from populations of 50,000 to populations of several million. Smaller MPOs are often housed within local government agencies, while larger MPOs may be independent agencies with broader regional planning powers. MPOs may also maintain other planning and environmental documents that are linked to their LRTPs. Some of these can also address climate change.

Southern California Association of Governments: 2008 Regional Transportation Plan (Adopted May 2008)

The Southern California Association of Government's (SCAG's) RTP includes a background section on the science of climate change and California's policies on climate change. The document also includes GHG emissions as a secondary performance measure for the achievement of environmental goals. (Emissions of criteria pollutants are the primary measure.)

SCAG does not state its transportation policies and strategies specifically in terms of climate change and reducing GHG emissions. Still, many of its strategies, including congestion relief, public transit promotion, and coordination of transportation and land use planning, do serve to reduce GHG emissions. The RTP mentions this connection in a subsequent section on mitigation of environmental impacts. For the next version of the RTP, SCAG will focus on refining techniques to estimate the impact of mitigation measures on GHG emissions.

In response to recent legal action under the California Environmental Quality Act (CEQA), SCAG's RTP Program Environmental Impact Report (PEIR) inventories the region's current GHG emission levels, outlines mitigation techniques, and considers the impact of various planning scenarios on GHG emissions reduction goals.

To reduce emissions that cause global warming, the PEIR proposes various mitigation strategies, from limiting the use of GHG emitting construction materials to increasing investment in non-motorized transportation. The PEIR also provides examples of how transportation planners may anticipate adaptation strategies for dealing with global warming, such as delineating floodplains and alluvial fan boundaries to prepare for hydrologic changes. The PEIR concludes "Improvements in air pollutant emission standards and increased use of alternative fuels would reduce GHG emissions. However, it is unlikely that mitigation measures would reduce GHG emission below existing levels (let alone to 1990 levels as required by AB 32) due to anticipated population growth."

SCAG also has a Regional Comprehensive Plan that addresses land use and housing, open space and habitat, water, energy, air quality, solid waste, transportation, security and emergency preparedness, and the economy. The Regional Comprehensive Plan assesses the potential impact of transportation policies and strategies on climate change.

Chicago Metropolitan Agency for Planning: 2030 Regional Transportation Plan (Updated June 2007)

The Chicago Metropolitan Agency for Planning (CMAP) states in its updated RTP, "We are concerned about transportation's role in the long-term sustainability of the natural environment as it relates to ecological concerns ranging from global climate change to natural beauty." Accordingly, one of the LRTP's goals is Employ Transportation to Sustain the Region's Vision and Values. Objectives under this goal include:

- "Promote transportation proposals that:
  - encourage reduced energy consumption, and
  - include elements that mitigate environmental problems including offsetting carbon emissions"

CMAP is also developing a new Regional Comprehensive Plan (RCP) that will integrate planning for transportation, land use, the natural environment, economic development, housing, and social systems. As part of the development of the RCP, CMAP hosted a Regional Climate Change Summit in December 2007. The agency also simultaneously issued a Regional Snapshot on Sustainability that addresses climate change as one of its key issues.

In developing the RCP, CMAP will analyze up to five different regional scenarios for their impacts on climate change. GHG emissions will be one performance indicator used to gauge the region's progress toward sustainability. CMAP has also actively sought input from stakeholders to determine the appropriate GHG reduction targets for the region and to discuss the extent to which land-use and transportation planning can help towards the eventual emissions goal. The outcome of these processes will undoubtedly feed into the next version of the LRTP.

Baltimore Regional Transportation Board: Transportation Outlook 2035 (Adopted November 2007)

The Baltimore Regional Transportation Board (BRTB) specifically mentions climate change within the Environmental Stewardship section of the transportation plan. BRTB recognizes that the plan does not directly address climate change and there are no established Federal standards for GHGs. Nevertheless the plan states that its goals of reducing vehicle emissions should help to reduce emissions of GHGs.

The specific strategies that will reduce GHG emissions include:

- Truck stop electrification
- Incident management
- Alternative fuel vehicle purchases
- Park-and-ride lot improvements
- Rideshare coordination
- Telework promotion

- Remote start devices for buses (which save idling time)

The plan also contains a policy to "Promote a sustainable environment by establishing policies that abate emissions from mobile sources, reduce energy consumption, reduce single occupant vehicles and the use of gasoline, and conserve and protect natural and cultural resources." The BRTB will support this policy with travel demand management programs and emission reduction projects. BRTB intends to fund strategies in three categories: technological, behavioral, and capital programs.

Central Lane Metropolitan Planning Organization 2007-2031 Regional Transportation Plan (Final Draft September 2007)

Eugene, Oregon's Central Lane Metropolitan Planning Organization (CLMPO) peripherally recognizes the threat of climate change in its LRTP.

While the LRTP does not mention climate change in the body of the document, many strategies in the document would help to reduce transportation GHG emissions. An appendix to the document mentions the connection:

"Given that budgets for transportation planning, construction, and maintenance are pinched already and concerns for global warming are on the rise, it would benefit the jurisdictions of the region to continue to support and enhance existing policies or strategies and develop new ones that reduce use of automobiles and encourage use of mass transit, carpooling, walking, bicycling, and telecommuting. Many of these strategies are discussed in the RTP…"

4.3. Barriers and Needs

DOTs and MPOs that wish to integrate climate change in their LRTP documents do not face any foreseeable statutory barriers. But some DOTs and MPOs are waiting on decisions or recommendations from state agencies or committees on how they should address climate change. Others see a need for greater involvement from federal or state government in climate change issues. Many agencies are wary of taking steps to change their planning documents before more direction from higher government levels is provided.

For example, the Florida Department of Transportation (FDOT) includes climate change in the goals of its LRTP, but the agency is not yet developing any strategies that explicitly address GHG emissions. FDOT is waiting for the recommendations of other state-level agencies and committees. Currently, the Florida Governor's Action Team on Energy and Climate Change is drafting recommendations that will include measures to reduce GHG emissions from transportation. There is also pending legislation in Florida that would require MPOs and local governments to consider climate change in their transportation plans. FDOT expects that these developments will help inform the next update of its statewide plan, which will likely begin in 2009. FDOT is also waiting for guidance from the governor on whether and how the agency's acclaimed Efficient Transportation Decision-Making process, which screens projects for environmental impacts, should incorporate GHG emissions.

Small MPOs in particular may benefit from higher level guidance on how and where to incorporate climate change in LRTPs. Small MPOs have fewer resources and less power to set policy precedents than do larger MPOs. The potential burden imposed by future climate change legislation at the state or federal levels is likely greater for small MPOs.

5. Quantification of GHG Impacts in Transportation Plans

Quantifying GHG emissions from transportation plans is a new area and one fraught with uncertainty. There are limitations in the ability of existing models to estimate the emissions generated by current and forecast transportation systems. Existing models may not adequately capture the potential reductions in emissions from certain strategies. While many agencies recognize that they will eventually need to estimate GHG emissions from plans and strategies, there are no standard tools or approaches yet.

Despite these uncertainties, DOTs and MPOs are taking some steps to quantify transportation GHG emissions now. Table 4 summarizes our findings on quantification requirements and processes among states, state DOTs, and MPOs. A number of agencies are using best available methods to quantify GHG emissions. Others are developing new methods and tools. The robustness and technical complexity of methods employed varies substantially. In some states, agencies are required—or will probably soon be required—to quantify GHG emissions. For those agencies instituting performance measures or targets related to GHG emissions, quantification is an indispensable step in the planning process.

Table 4: State and MPO Quantification of GHG Emissions from Transportation Plans

State requirement for quantification	Current	New York	
	Considering*	California, Oregon, Washington	
DOT and MPO quantification	Completed	Albany, NY MPO; Sacramento MPO; San Diego MPO	Southern California MPO; Maine DOT
	Forthcoming	Chicago MPO; Missoula MPO; Philadelphia MPO	Portland MPO; San Francisco MPO; Seattle MPO

* These states have proposed requirements in official documents.

For those DOTs and MPOs pursuing quantification of GHG emissions, quantification can include one or more of several elements. Transportation GHG inventories document historic (and possibly projected) GHG emissions. Inventories are often created as part of statewide climate action plans. Climate action plans often quantify changes in emissions from mitigation strategies as well. Some metropolitan regions are now also creating GHG inventories and climate action plans. Within the transportation planning process, emissions resulting from LRTP alternative investment packages can be calculated. Emissions resulting from specific projects or bundles of projects can also be calculated.

5.1. Transportation Plan GHG Impacts – Current Practice

Many transportation agencies are not currently attempting to quantify GHG emissions from transportation plans. Those agencies that are working to quantify emissions are taking different approaches, commensurate with their levels of resources and expertise in the area.

Metropolitan Transportation Commission

The Metropolitan Transportation Commission (MTC), the MPO for the San Francisco Bay Area, will adopt performance targets for GHG emissions in the next version of its RTP, a draft of which is due for release in December 2008. The preliminary target is to reduce CO2 emissions 40% below 1990 levels by 2035. A second preliminary target is to reduce VMT per capita by 10 percent by 2035. The planning process has established these targets upfront. Proposed packages of investments are being analyzed for their ability to meet these and other targets. Packages include freeway investment with modest efficiency improvements, a high-occupancy toll (HOT) network with expanded express bus service, an expansion of rail transit, a comprehensive road-pricing policy, and a land-use strategy based on smart growth principles. The approach to analysis differs from past planning efforts, in which system performance was assessed only after packages of investments were proposed. The plan's preliminary CO2 targets are based on California's state-level mandates, including AB32.[1]

MTC is currently evaluating the CO2 impacts of individual highway and transit projects. This analysis will feed into a performance comparison of projects. A project-level GHG analysis raises some key issues such as where to set the boundaries for analysis and how to account for potential mode shift.

Puget Sound Regional Council

The Puget Sound Regional Council, the MPO for the Seattle area, is using the U.S. EPA's Motor Vehicle Emission Simulator (MOVES) model to do a regional level analysis of GHG emissions in its LRTP. As part of the analysis, PSRC is helping to pilot an updated version of MOVES that includes improved sensitivity of emissions to vehicle speeds. PSRC also has received a grant from FHWA to make improvements in their travel demand model. The improvements will capture the impacts of factors like mode choice and the cost of driving. Outputs from the model will feed into the GHG emissions analysis.

Missoula County

Missoula County serves as the MPO for Missoula, Montana and is currently updating its LRTP. The update process includes a regional land use and transportation visioning exercise called Envision Missoula. Missoula plans to provide a basic estimate of CO2 emissions from the plan, probably using a simple VMT multiplier applied to the outputs of the travel demand model. The agency would be open to receiving guidance on other appropriate techniques for GHG emission quantification.

Capital District Transportation Committee

Albany's Capital District Transportation Committee (CDTC) incorporates analysis of GHG emissions in its planning process in two ways. First, CDTC applies a "full cost analysis", including analysis of global warming costs, to major system decisions. Full cost analysis is also used to evaluate candidate TIP

---

[1] Droettboom, Ted and Christy Riviere, "Breaking with the Past." Planning. May 2008. American Planning Association, Chicago, IL. 33-34.

projects, when applicable. Second, in compliance with New York State requirements, the agency estimates the GHG emissions resulting from its LRTP. (As described in Section 3.2, New York requires MPOs to estimate the energy and CO2 emissions from their long range transportation plans and also from their transportation improvement programs.) CDTC modified is post-processor to produce GHG emissions specific to year, functional classification, and operating speed.

CDTC has taken an innovative approach to the use of their travel demand modeling. The region has been proactive in encouraging concentrated, sustainable development patterns, and has a focused interest in establishing linkages between policy setting and environmental responsibility. CDTC forecast a 15% reduction in trip generation per household based on a range of policies and principles, such as urban reinvestment, transit oriented development, and bus rapid transit. CDTC believes that travel demand forecasts are partly a self-fulfilling prophecy, since forecasts are used in project development and design. They are advocating no new major highway construction, and that any widening involve managed lanes and HOV. Based on these assumptions, CDTC has calculated impacts on GHG emissions, fuel consumption, and air quality emissions. The New York State Department of Environmental Conservation (NYSDEC), which reviews CDTC's air quality and conformity reports, has not accepted the growth forecast, saying it understates future problems. Both CDTC and NYSDOT are currently advocating for the approach.

Sacramento Area Council of Governments

The Sacramento Area Council of Governments (SACOG) used the results of a report from the California Energy Commission (CEC) to estimate the impact of its LRTP on CO2 emissions. The report, entitled "Effect of Land Use Choices on Transportation Fuel Demand", estimated the fuel savings associated with SACOG's LRTP. In addition, SACOG is working with several modeling applications to analyze the impacts of different transportation and land use scenarios. SacSim, the agency's new travel demand forecasting model, is the first regional model to use individual land parcels as the level of input data.[2] The agency is working to create new linkages between its travel demand model and models that include land use and economic forecasts. SACOG is also working to improve its model of commercial vehicle travel behavior.

SACOG has estimated that its share of the statewide GHG reduction goal for the regional transportation/smart growth sector is 1 million metric tons CO2-equivalent (MMtCO2e).[3] The EIR used this share to determine the impact significance of the LRTP on climate change.

5.2. Transportation GHG Inventories – Current Practice

A majority of states have developed statewide GHG inventories as a reporting tool to track annual emissions and inform policy development. Statewide GHG inventories typically provide detailed information on the volumes, types, and sources of emissions and allow for comparisons of emissions over time and across source types.

---

[2] SACOG, "Comments on Placer Vineyards Specific Plan, Second Partially Recirculated Revised Draft EIR." May 15, 2007

[3] The statewide total (18 MMtCO2e) is as determined by the Climate Action Team in the report "Climate Action Team Proposed Early Actions to Mitigate Climate Change in California, Draft for Public Review" (CalEPA, April 2007)

Some MPOs are also involved in regional GHG estimation efforts as part of a broader regional climate change planning process. Development of a regional GHG inventory is a relatively new concept; two regions, the Philadelphia and Washington, DC urban areas, are currently engaged in this process with support from EPA. Transportation emissions are one component of GHG inventories.

Delaware Valley Regional Planning Commission, Philadelphia PA

The Delaware Valley Regional Planning Commission (DVRPC) is in the process of preparing a regional GHG inventory for 2005, as well as projected GHG emissions for 2035. Among the sources to be included in this inventory are emissions from on-road transportation, which are expected to be one of the region's primary sources of GHG emissions. CO_2, CH_4, and N_2O emissions will be calculated using modeled estimates of annual average daily vehicle miles traveled (VMT) by vehicle type and road class. Per mile emissions factors will be applied to the VMT totals.

The VMT estimates will be generated by DVRPC's regional transportation model, which is used to support the region's long range transportation planning and air quality conformity analysis process. The regional transportation model will also be used to generate GHG emissions estimates for various transportation plan alternatives. For both the 2005 estimates and 2035 projections, the VMT will be reported for each county. Thus the region's transportation emissions can be allocated to the county level. Preliminary estimates are expected from this effort by summer 2008, with final results by November 2008.

Metropolitan Washington Council of Governments, Washington DC

A regional inventory of CO_2 emissions from transportation was developed by the Metropolitan Washington Council of Governments (MWCOG) while calculating 1990 and 2005 emissions of criteria pollutants for air quality conformity purposes. CO_2 estimates from mobile sources were calculated using data and forecasts of vehicle miles of travel (VMT) by vehicle type from the air quality conformity analysis. Emission factors were modeled using MOBILE6 and travel patterns in the COG region on network and local roadways. The inventoried emissions included auto access to transit and diesel transit and school buses. Emissions forecasts to 2030 were developed using assumptions in the MOBILE6 model (such as changes in fuel mix over time) and the COG's travel forecasting model.

MWCOG has proposed two GHG emission reduction scenarios for development. The first scenario reflects the current procedures of the LRTP and uses goals that are within reach fiscally and administratively, but that improve the conditions of the 2030 baseline. The second scenario examines how new long-term goals could be achieved using various combinations of interventions, including improved fuel efficiency, alternative fuels, and reducing vehicle travel. The first step in developing this scenario is identifying a CO_2 emission reduction goal. MWCOG has noted that one of the main benefits from establishing a reduction goal is that many of the strategies used to reduce CO_2 emissions will also provide ancillary transportation, environmental, health, and economic benefits, such as reduced congestion and fuel consumption. The COG Climate Change Steering Committee has discussed a proposed regional goal of reducing overall regional CO_2 emissions by 70-80% below 2005 levels by 2050.

5.3. GHG Impacts from Construction – Current Practice

Most agencies are focusing on estimating the emissions associated with the *use* of transportation facilities. A few agencies are also attempting to quantify GHG emissions associated with *construction and maintenance* of transportation facilities.

Metropolitan Transportation Commission

At the request of the U.S. Environmental Protection Agency (EPA), MTC is considering how to quantify emissions from construction of transportation facilities. The agency plans to estimate CO_2 emissions from construction in the Environmental Impact Report (EIR) for its RTP. No methodology has yet been determined. One possible approach is to use outputs from the Sacramento Air Quality Management District's Roadway Construction Emissions Model, along with CO_2 emissions factors from another source, to calculate emissions.

Capital District Transportation Committee

Albany's CDTC quantified indirect energy use associated with their TIP. The NYSDOT methodology for calculating GHG emissions describes indirect energy as the energy required to construct and maintain transportation facilities. The CDTC estimate of this energy use was 931 billion Btu over the 5-year analysis period of the TIP. However, indirect impacts could not be estimated without a detailed and very time-consuming analysis of specific project schedules in each region. CDTC used NYSDOT's methodology to perform the calculations, and only included energy for road construction, and only addressed projects listed in the TIP, not additional projects proposed in the long range transportation plan.

5.4. Barriers and Needs

Quantification of GHG emissions is one of the most challenging aspects of integrating climate change into transportation planning. There is room for improvement across the board in inventory techniques and techniques for estimating the impact of policies and strategies. Guidance is particularly needed in this area.

Most state-level GHG inventories for transportation have some major limitations for transportation policy analysis because they are not presented in sufficient detail to assess emissions reduction strategies. Inventories of transportation GHG emissions are typically developed by fuel type based on fuel sales data, while strategy analysis requires estimate of emissions for individual modes, vehicle types, and geographic areas. In addition, statewide inventories do not report emissions at the regional level, and therefore are of limited use for MPO planning and strategy analysis.

Consistency of transportation data is a problem for GHG inventories. The data on fuel sales that are used to calculate statewide inventories do not always match with data on vehicle miles traveled (VMT). Both cross-border travel and poor information on state fleets and fuel economy contribute to this discrepancy. These issues are magnified at smaller geographical levels. In developing regional transportation GHG inventories, MPOs will rely heavily on local VMT estimates and perhaps information on local travel conditions. Such local inventories are very likely to be inconsistent with state-level inventories. If and when regions are required to meet certain VMT or transportation GHG reduction goals, state and regional inventories would provide conflicting bases for performance measurement. The development of reliable and consistent transportation GHG inventories at the regional scale is therefore important.

The appropriate level at which to quantify emissions in transportation plans remains an area of uncertainty for transportation agencies. Regional level analyses compare GHG emissions between broad packages of modal and development strategies. Many agencies argue that quantitative analyses are not useful at finer grains, because GHG emissions are essentially determined at the regional level. Other agencies are proceeding with analyses down to the project level.

Guidance on appropriate quantification techniques for various agencies and planning components is needed. There is a high potential for disagreement over appropriate quantification methods at present. The experience of Albany's CDTC illustrates. While the current atmosphere allows agencies to be innovative in their approaches to quantification, agencies with fewer resources to develop and test quantification methods are at a significant disadvantage.

The current emissions models are another barrier to quantification of GHG emissions. While EPA's MOBILE6 model can produce CO2 emissions estimates, the CO2 emission factors do not vary with vehicle speed or driving cycle. For this reason, MOBILE6 is inappropriate for any kind of detailed analysis of transportation plan or project-level emissions, which are likely to involve changes to congestion levels and speeds. EPA's new MOVES2004 model does have GHG emission factors that vary with speed and driving cycle. However, rather than using the MOVES2004 current interface, users must manually modify the MOVES database and create specific driving cycles in order to take advantage of this information. This presents a challenge to many users of the model, and also makes it cumbersome to analyze a large number of scenarios.

As state and federal agencies consider regulations and guidance that involves quantification of transportation GHG emissions, they should keep in mind particularly the needs of smaller MPOs. For example, Eugene's CLMPO, with just four full time staff members, is concerned about having expectations or standards for climate change planning set by the larger MPOs in the state. It is therefore important that small MPOs such as CLMPO receive explicit consideration in the development of state or federal planning requirements. For example, unfunded state mandates can be particularly burdensome for small MPOs that rely exclusively on federal funding. CLMPO is actively participating in discussions on climate change at the state level in order to prevent any oversight of the needs and constraints of small MPOs.

6. GHG Mitigation Strategies in Transportation Planning

Most transportation agencies have not yet developed or analyzed strategies explicitly to reduce GHG emissions, although many agencies have ideas about promising strategies. In most states, the only GHG analyses of transportation strategies are those being done for state climate action plans.

Ultimately GHG mitigation strategies will be indispensable to reducing the impact of transportation on climate change. MPOs and DOTs can begin to think about what strategies are available to them for immediate and future implementation.

6.1. Land Use Integration

Many agencies see integrated transportation and land use planning as the most effective strategy to reduce transportation GHG emissions in the long term. But MPOs and DOTs do not have the legal authority on their own to conduct effective integrated planning. Some agencies see the need for better planning processes, while others are making use of existing powers to encourage better land use and transportation planning.

Metropolitan Transportation Commission

In the San Francisco Bay Area, MTC projects that aggressive smart growth policies will be a key component in efforts to attain GHG emission reduction targets. A system-level analysis of policy and investment alternatives projected that smart growth policies would reduce CO_2 emissions from transportation by 6-8%. In order to achieve the proposed plan target, additional strategies would be needed including investments in transit, HOT lanes, and freeway system performance technologies, pricing policies, and increases in telecommuting and vehicle fuel efficiency.[4] Table 5 below shows the results of this analysis.

---

[4] Metropolitan Transportation Commission, "Travel Forecasts for the San Francisco Bay Area 2009 Regional Transportation Plan Vision 2035 Analysis: Data Summary." November 2007. Summary Exhibit 1. Available at: http://www.mtc.ca.gov/planning/2035_plan/tech_data_summary_report.pdf.

Table 5: Comparison of MTC Plan Alternatives: 2035 CO2 Emissions (Thousand Tons/Day)

	Infrastructure Packages			
	No New Investments	Freeway Performance	HOT & Local/ Express Bus	Regional Rail & Ferry
Baseline Emissions (No Policy Changes)	101.4	92.4	97.0	99.1
Reductions from Policy Packages				
Pricing Sensitivity	-8%	-6%	-8%	-8%
Land Use Sensitivity	-8%	-6%	-7%	-7%
Combined Pricing & Land Use	-14%	-11%	-13%	-14%
Combined Pricing, Land Use, and Telecommuting	n/a	-14%	-17%	n/a
Combined Pricing, Land Use, Telecommuting and Fuel Efficiency	n/a	n/a	-46%	n/a

Capital District Transportation Committee

The Albany Capital District Transportation Committee (CDTC) incorporates greenhouse gas emissions and energy efficiency in its LRTP, New Visions 2030. One of the plan's four broad themes is the linkage of transportation and land use, recognizing the impact that better transportation and land use planning can have to reduce GHG emissions. To implement the strategy, CDTC makes a concerted effort to partner with local communities to bring them on board in the planning process. To this end, CDTC sponsors the Community and Transportation Linkage Planning Program, which gives matching grants worth $30,000-50,000 to local communities to hire a consultant to do corridor transportation and land use studies. Within these projects, visualization techniques allow the public to understand and approve the principles of linking transportation and land use. CDTC has distributed nearly $3 million through this program over the last few years, to complete a total of 53 linkage studies. The program is popular with local communities.

6.2. Pricing and Alternative Modes

Other strategies to reduce GHG emissions, such as road pricing and improved transit, face a variety of political and funding challenges. To date, there has been relatively little action by MPOs and DOTs to analyze or pursue these types of strategies in the name of GHG reduction, although many expect this to change in the near future.

Portland Metro

In addition to land use strategies, Portland Metro sees improvements to system efficiency as a valuable strategy to reduce GHG emissions. Specifically, Metro is encouraging the use of congestion pricing on roads in the Portland area. Currently public sentiment is largely opposed to congestion pricing, and elected officials are therefore reluctant to support it. There is also some public apathy towards Metro's employer trip reduction program. Residents and employers do not always see the connection between the program and broader transportation goals. In addition, Metro is facing a major funding shortfall for transit expansion projects. The currently identified funding gap is $7 million.

Metropolitan Transportation Commission

MTC's analysis (described in Section 6.1) highlights congestion pricing as an important strategy to reduce GHG emissions. The analysis of policy and investment alternatives projected that pricing policies would reduce CO2 emissions from transportation by 11-14%.[5] (See Table 5 above.) But implementing congestion pricing is politically challenging. MTC would need the support of state legislation to implement widespread pricing. The recent controversy over a proposed tolling scheme on Doyle Drive in San Francisco illustrates the challenges. The proposal has met with strong opposition from some local stakeholders.

6.3. GHG Benefits of Existing Strategies

Nearly all transportation agencies already have strategies in place that reduce GHG emissions, such as TDM, transit promotion, and congestion reduction. Many agencies have not explicitly recognized or evaluated the potential of these existing strategies to contribute to mitigation of GHG emissions, but agencies can use climate change to promote existing strategies.

Central Lane Metropolitan Planning Organization

Eugene's CLMPO programs over a half million dollars a year for demand management programs. These programs have been in place since before climate change became a major issue in transportation planning. The current adopted version of the plan makes only a peripheral reference to the ability of such programs to reduce GHG emissions; but that plan, adopted in September 2007, is the result of an update cycle that begin in 2003. Recently CLMPO has begun to devote more attention to climate change issues. The MPO is now thinking more about how such strategies will help to reduce GHG emissions.

Capital District Transportation Committee

Albany's Capital District Transportation Committee (CDTC) considers the new attention on climate change an opportunity to reinforce the type of sustainable planning that they have been advocating for some time. Climate change now gives them further rationale and justification for the agency's traditional approaches to transportation planning, which are centered around meeting the region's transportation needs in a cost effective manner while also promoting safety, enhancing the environment, building strong communities, and improving the overall quality of life. While there is an agency-wide effort on climate change, CDTC doesn't find a need to change established planning process to integrate climate change considerations. For example, CDTC's TIP has already been customized to encourage projects that favor transit and TDM, and could thus be considered a climate change mitigation strategy.

6.4. Barriers and Needs

GHG mitigation strategies face both funding and political barriers. MPOs and DOTs typically do not have the authority on their own to implement some of the most promising strategies, including integrated land use and transportation planning and road pricing. Without support from key partners, such strategies may not be viable. For strategies that require significant capital investments, particularly transit strategies, the support of key partners is also indispensable for securing funding.

---

[5] MTC November 2007, Summary Exhibit 1.

Many states have had initial success with developing broadly accepted GHG reduction strategies through multi-agency work groups. By involving diverse stakeholders in the drafting of climate action plans, these work groups develop politically viable strategies to reduce emissions. Some regions are just beginning work on their own multi-stakeholder climate action plans. These plans may produce better strategies for implementation by MPOs.

Guidance is also needed from higher levels on appropriate and cost effective GHG reduction strategies for urban areas of various sizes. In some cases, support from state government may empower MPOs to pursue strategies that would otherwise be difficult to achieve.

7. Climate Change Adaptation in Transportation Planning

The threats that climate change poses to transportation systems—including flooding, changes in average temperatures, and extreme weather events—are clear. But MPOs and DOTs have little if any information on precisely what impacts they can expect, where, and in what time frames. As a result, agencies are largely not acting to adapt the transportation system to climate change, or are waiting for further guidance on the topic. But agencies can still take preliminary steps to study the issues.

Metropolitan Transportation Commission

The Metropolitan Transportation Commission is in the very early stages of a research study on the potential impacts of climate change on Bay Area transportation infrastructure. The study will examine the risk from sea level rise, coupled with storm surge. MTC is conducting this study in coordination with a regional partner, the Bay Conservation and Development Commission. It is not yet clear how the study will feed into long range planning processes.

Puget Sound Regional Council

Puget Sound Regional Council is not currently taking action to adapt the transportation system to a changing climate. Deciding how to incorporate adaptation into the LRTP is a particular challenge. Like many MPOs, PSRC is an agency that plans for and coordinates the efforts of local jurisdictions; the agency has no implementing authority. It is not clear what the role of the agency should be in establishing protective measures.

Still, PSRC specifically recognizes the need to consider adaptation to climate change in the LRTP including impacts on the transportation system such as:[6]

- Accelerated pavement deterioration
- Flooded roadways
- Bridge damage
- Increased maintenance
- Increased stormwater, drainage issues

New York State DOT

New York State DOT (NYSDOT) in interested in thinking about adaptation, but has yet to focus on it within transportation planning. They have created a working group for climate change issues, which includes MPOs and the NYSDEC, that is considering a number of issues and conducting an assessment to decide what issues to focus on. Adaptation is being considered within this group. If NYSDOT had the

---

[6] "Transportation Planning and Climate Change," Presentation by Kelly McGourty, PSRC, to the WSDOT Transportation Planning Symposium, November 14, 2007. Available at:
www.wsdot.wa.gov/NR/rdonlyres/92A9B1E2-6046-486E-B468-06275648830F/0/IClimateChangeMcGourty.pdf

resources, the agency would like to conduct a study on regional sea level rise and climatic changes, similar to the U.S. Climate Change Science Program's recently released Gulf Coast Study.[7]

---

[7] U.S. Climate Change Science Program (USCCSP), Impacts of Climate Change and Variability on Transportation Systems and Infrastructure: Gulf Coast Study, Phase I. Synthesis and Assessment Product 4.7, March 2008. Available at: http://www.climatescience.gov/Library/sap/sap4-7/final-report/.

8. Institutional Relationships

Because of concern about climate change, MPOs and DOTs are participating in or leading a number of new inter-agency initiatives. Agencies recognize the need for cross-agency collaboration on climate change issues. Working together, agencies can transcend jurisdictional boundaries, pool resources, and share expertise. Many multi-agency groups are defining policy at the state and regional levels.

DOTs and MPOs that are not already involved in such groups may find them to be effective vehicles to develop mitigation strategies and quantification tools as well as determine common planning and analysis methods. Table 6 lists some agencies already involved in such groups.

Table 6: MPOs and DOTs Involved in a Multi-agency Climate Change Initiative or Plan

MPOs	DOTs
- Chicago Region	- Arizona
- Denver Region	- California
- Houston-Galveston Region	- Connecticut
- San Diego Region	- Florida
- San Francisco Bay Area	- Maine
- Seattle Region	- Maryland
- Washington, DC Region	- Massachusetts
	- New York
	- Oregon
	- Washington

Note: This table is not a comprehensive list; it reflects only state DOTs and MPOs reviewed as part of this study.

8.1. Current Practices

Oregon DOT

The Oregon Department of Transportation is actively involved in several executive level efforts on climate change. A representative from ODOT provides input to Oregon's Global Warming Commission, an executive level advisory body. ODOT also has a representative in a working group of the Governor's Transportation Vision Committee, which is considering the need for transportation funding strategies that help to reduce GHG emissions. ODOT sees these committees as integral to the development of strategies to reduce GHG emissions through transportation planning and expects any new strategies will be developed in concert with these groups.

Chicago Metropolitan Agency for Planning

At Chicago's CMAP, the Environmental Committee is at the center of the planning process, including planning for climate change. CMAP is dedicated to engaging their environmental partners as an integral part of the regional planning process, not as an afterthought. When the new director took the lead of CMAP, he invited the environmental community to provide input to the transportation planning process.

The partnerships have grown easily since then. Since CMAP is a new agency, many environmental stakeholders felt they had an opportunity to become involved in the early phases, and have remained engaged in the planning process since then.

California DOT

Caltrans participates in California's Climate Action Team (CAT), which was created in response to an executive order from Governor Schwarzenegger. The group is headed by the California Environmental Protection Agency and includes representatives from other state level agencies with jurisdiction over environmental affairs and natural resources. The CAT plays an important role in developing strategies for the state to comply with the emissions reduction goal established by AB32. In addition, Caltrans supports the California Air Resources Board (CARB) in its role as developer of regulations and standards to reduce GHG emissions.

Washington DOT

Washington DOT (WSDOT) is heavily involved in multi-agency climate change activities, due in part to state legislation and executive orders. In response to Executive Order 07-02 and SB 6001, Washington completed a statewide climate action plan in early 2008, which resulted in more than 50 policy recommendations including 12 focused on reducing transportation GHGs. WSDOT was a member of the Transportation Technical Working Group that guided the policy development process.

In March 2008, Washington Governor Gregoire signed climate change framework legislation HB 2815, which includes (among other things), a requirement to reduce light duty vehicle per capita VMT 18% by 2020, 30% by 2035, and 50% by 2050. The state has formed a Climate Action Team (CAT), which includes Washington's Transportation Secretary. The CAT is being advised by sector-specific Implementation Working Groups, each comprising a diverse set of stakeholders and charged with developing the specific implementation steps needed to achieve the reductions in the climate plan. A WSDOT representative co-chairs the Transportation Implementation Working Group. The group is currently developing recommendations for specific actions that can help achieve the VMT reduction goals laid out in HB 2815

8.2. Barriers and Needs

While agency partners appear very committed to climate change action, and there is a high degree of cooperation, many MPOs and DOTs find that the sheer volume of activity on climate change and the pace of development make coordination difficult. Broader sharing of best practices for such inter-agency efforts may benefit both MPOs and DOTs

9. Conclusion

Many DOTs and MPOs are beginning to incorporate climate change issues into their transportation planning processes. Within a few years, it is likely that virtually all new transportation plans will include explicit reference to the effects of transportation on climate change and the role of transportation in mitigating these effects. Many plans will also likely address the threats that climate change poses to the transportation system and potential adaptive responses.

While they do not explicitly require the inclusion of climate change considerations, the current federal transportation planning regulations include a number of requirements that generally align with climate change mitigation and adaptation. This occurs most directly in metropolitan transportation planning factor that requires that plans "protect and enhance the environment [and] promote energy conservation".

The current practice for incorporating climate change into transportation planning varies widely by agency. Climate change can appear in the vision, goals, policies, strategies, trends and challenges, and performance measures of LRTPs. Some plans merely recognize that climate change is an issue that relates to transportation and begin to point out the relevance of existing plans and strategies to climate change. Other plans make climate change more central to their goals and policies.

Absent any federal action, the treatment of climate change in transportation planning is likely to continue to vary depending on the interests and concerns of local stakeholders, the size of agencies and their capacity to address climate change, and the vulnerabilities specific to regions and their transportation systems. A number of agencies DOTs and MPOs are waiting on decisions or recommendations from state agencies or committees on how they should address climate change. Others see a need for greater involvement from federal or state government in climate change issues. Many agencies are wary of taking steps to change their planning process before more direction from higher government levels is provided.

Small MPOs in particular may benefit from higher level guidance on how and where to incorporate climate change in LRTPs. Small MPOs have fewer resources and less power to set policy precedents than do larger MPOs. The potential burden imposed by future climate change regulations at the state or federal levels is likely greater for small MPOs.

The quantification of GHG emissions in the transportation planning process is a new challenge for transportation agencies. While the estimation of mobile source CO_2 emissions is conceptually simpler than the estimation of criteria pollutant emissions that most transportation agencies already do, there are some unique challenges with the quantification of GHGs, including the following:

- There are often inconsistencies between the transportation components of a state-level GHG inventory and forecast (based on fuel sales data) and a metropolitan-level GHG inventory and forecast (based on VMT estimates from a travel demand model). As state-level climate change initiatives filter down to the MPO level, these inconsistencies will need to be reconciled.

- There are uncertainties regarding the appropriate geographic scale for estimation of project-level GHG emissions. Regional level analyses compare GHG emissions between broad packages of modal and land development strategies. Some agencies argue that quantitative analyses are not useful at finer grains, because GHG emissions are essentially determined at the regional level. Other agencies are proceeding with analyses down to the project level.

- The current EPA emissions models are not well-suited for analysis of the GHG impacts of transportation plans. The CO_2 emission factors in MOBILE6 do not vary with vehicle speed or driving cycle, and thus are inappropriate for any kind of detailed analysis of transportation plan or project-level emissions, which are likely to involve changes to congestion levels and speeds. The

MOVES2004 model does have GHG emission factors that vary with speed and driving cycle, but rather than using the MOVES2004 current interface, users must manually modify the MOVES database and create specific driving cycles in order to take advantage of this information. This presents a challenge to many users of the model, and also makes it cumbersome to analyze a large number of scenarios.

- Given the importance of a life-cycle approach to GHG emissions analysis, there is uncertainty regarding the need to estimate emissions resulting from transportation system construction and maintenance, as opposed to system use. Methods to estimate construction GHG emissions are poorly developed.

Many transportation agencies are anticipating the need to develop and quantify the benefits of strategies to reduce GHG emissions. A number of DOTs and MPOs have been involved in this exercise through their participation in state climate action plans. A few MPOs are taking steps to incorporate GHG mitigation into their planning, prompted by state mandates. There is concern among some transportation agencies that many of the most effective mitigation strategies are outside their sphere of direct influence (such as vehicle fuel efficiency, alternative fuels, and land use), while other potentially effective strategies (such as widespread use of roadway pricing) may be politically difficult.

Most transportation agencies are not currently seeking to incorporate climate change adaptation measures into long range planning. While there is general recognition of the threat that climate change poses to transportation infrastructure, agencies feel that significant impacts are at least several decades away, so there is little sense of urgency. In addition, the large uncertainty in the location and magnitude of impacts makes agencies reluctant to take major action on adaptation, given the multitude of other pressing demands for DOTs and their funding limitations. Over the next several years, as more sea level rise studies are completed and scientists improve the precision of climate change forecasts, adaptive responses are likely to be more substantially incorporated into long range planning.

Appendix A: List of Interviewees

New York State DOT

Karen Rae and Mary Ivey

Massachusetts Executive Office of Transportation

Kate Fichter

Capital District Transportation Committee (Albany, NY)

Chris O'Neill

Chicago Metropolitan Agency for Planning

Randy Blankenhorn

Florida DOT

Kathleen Neill

Metropolitan Transportation Commission

Harold Brazil

Puget Sound Regional Council

Kelly McGourty

Missoula County, Montana

Mike Kress

Metro (Portland)

Kim Ellis

Central Lane Metropolitan Planning Organization

Susan Payne, Paul Thompson

Oregon DOT

Damon Fordham, Barbara Frafer

Sacramento Area Council of Governments

Gordon Garry

Appendix B: Opportunities for Integrating Climate Change Considerations into Federal Transportation Planning Regulations

Title 23 USC 134 – Metropolitan Transportation Planning (FHWA)			
Section	Aspects	Language	Relation to Climate Change
§134(a)	Policy	It is in the national interest to— (1) encourage and promote the safe and efficient management, operation, and development of surface transportation systems that will serve the mobility needs of people and freight and foster economic growth and development within and between States and urbanized areas, while **minimizing transportation-related fuel consumption and air pollution** through metropolitan and statewide transportation planning processes identified in this chapter;	*Provides a link to GHG mitigation through emphasis on minimizing fuel consumption, since GHG emissions from transportation are largely correlated with fuel consumption, and air pollution.*
§134(c)	General Requirements	(2) Contents.— The plans and TIPs for each metropolitan area shall provide for the development and integrated management and operation of transportation systems and facilities (including accessible pedestrian walkways and bicycle transportation facilities) that will function as an intermodal transportation system for the metropolitan planning area and as an integral part of an intermodal transportation system for the State and the United States.	*To ensure an integrated transportation system to serve the country, MPOs will need to consider the implications of climate change (such as sea level rise) on their infrastructure to ensure effective connectivity is preserved. Additionally, emphasis on non-motorized transportation and could also facilitate climate change mitigation strategies.*
§134(g)	MPO Consultation in Plan and TIP Coordination	(3) Relationship with other planning officials— The Secretary shall encourage each metropolitan planning organization to **consult with officials responsible for other types of planning activities that are affected by transportation** in the area (including State and local **planned growth, economic development, environmental protection, airport operations, and freight movements**) or to coordinate its planning process, to the maximum extent practicable, with such planning activities. Under the metropolitan planning process, transportation plans and TIPs shall be **developed with due consideration of other related planning activities** within the metropolitan area…	*MPOs should consider, as part of the consultation requirement, climate action planning activities going on within their State or region, as well as local government plans or policies that may consider climate change.*

§134(h)	Scope of Planning Process	(1) In general.— The metropolitan planning process for a metropolitan planning area under this section shall provide for consideration of projects and strategies that will— (A) support the economic vitality of the metropolitan area, especially by **enabling global competitiveness, productivity, and efficiency**; (B) increase the **safety** of the transportation system for motorized and nonmotorized users; (C) increase the **security** of the transportation system for motorized and nonmotorized users; (D) increase the **accessibility and mobility** of people and for freight; (E) **protect and enhance the environment, promote energy conservation, improve the quality of life, and promote consistency between transportation improvements and State and local planned growth and economic development patterns**; (F) enhance the **integration and connectivity** of the transportation system, across and between modes, for people and freight; (G) promote efficient system management and operation; and (H) emphasize the **preservation** of the existing transportation system.		*Requirements for strategies to address safety (B), security (C), accessibility (D), connectivity (F), and preservation (H) will require some MPOs to consider projected impacts of climate change, including sea level rise, on infrastructure. Adaptation strategies will need to be implemented to ensure continued connectivity and accessibility, as well as to promote security of the system, ensure the safety of the system for users, and to support global competitiveness and efficiency (A).* *Requirement of strategies in (E) provides a link to GHG mitigation through emphasis on energy conservation (since GHG emissions from transportation are largely correlated with energy consumption) and consideration of environmental protection.*
§134(i)	Development of Transportation Plan	(2) Transportation plan.— A transportation plan under this section shall be in a form that the Secretary determines to be appropriate and shall contain, at a minimum, the following: (A) Identification of transportation facilities.— An identification of transportation facilities (including major roadways, transit, multimodal and intermodal facilities, and intermodal connectors) that should function as an integrated metropolitan transportation system, giving emphasis to those facilities that serve important national and regional transportation functions. In formulating the transportation plan, the metropolitan planning organization shall consider factors described in subsection (h) as such factors relate to a 20-year forecast period.		*In considering ensuring an integrated transportation system (2A) and preservation of the projected and existing system (E), adaptation strategies could be required or encouraged, since some MPOs will need to consider the implications of climate change (such as sea level rise) on their infrastructure to ensure effective connectivity is preserved. Temperature swings resulting from climate change are also expected to cause accelerated aging on infrastructure.* *Mitigations activities are specifically required within the statewide plan (2B); climate change mitigation strategies could be directly linked to this requirement.* *Management and operations strategies (2D) can often be considered climate change mitigation strategies, if they improve system performance and reduce emissions. Similarly, transit*

		(B) Mitigation activities.—

(i) In general.— A long-range transportation plan shall include a discussion of types of **potential environmental mitigation activities** and potential areas to carry out these activities, including activities that may have the greatest potential to restore and maintain the environmental functions affected by the plan.

(ii) Consultation.— The discussion shall be **developed in consultation with Federal, State, and tribal wildlife, land management, and regulatory agencies**.

(D) Operational and management strategies.— Operational and management strategies to **improve the performance of existing transportation facilities** to relieve vehicular congestion and maximize the safety and mobility of people and goods.

(E) Capital investment and other strategies.— Capital investment and other strategies to **preserve the existing and projected future metropolitan transportation infrastructure** and provide for multimodal capacity increases based on regional priorities and needs.

(F) Transportation and transit enhancement activities.— Proposed transportation and **transit enhancement** activities.

3) **Coordination with clean air act agencies.**— In metropolitan areas which are in nonattainment for ozone or carbon monoxide under the Clean Air Act, the metropolitan planning organization shall coordinate the development of a transportation plan with the process for development of the **transportation control measures of the State implementation plan** required by the Clean Air Act.

(4) Consultation.—

(A) In general.— In each metropolitan area, the metropolitan planning organization shall **consult, as appropriate, with State and local agencies responsible for land use management, natural resources, environmental protection, conservation,** and historic preservation concerning the development of a long-range transportation plan.

(B) Issues.— The consultation shall involve, as appropriate— | *enhancements (2F) can potentially serve as climate change mitigation strategies.*

Additionally, direct linkages are possible in consultation with agencies responsible for or involved with climate action planning, including environmental and land use agencies that might also be incorporating climate change considerations – both mitigation and adaptation – into their planning or programs (2B(ii), 3 and 4).

There is an opportunity to add new language to consultation, under "issues" (4B) to specifically address climate change through consideration of climate change plans in addition to the specified conservation plans. |

		(i) comparison of transportation plans with State conservation plans or maps, if available; or (ii) comparison of transportation plans to inventories of natural or historic resources, if available.	
§134(k)	Transportation Management Areas	(3) Congestion management process.— Within a metropolitan planning area serving a transportation management area, the transportation planning process under this section shall address congestion management through a process that provides for effective management and operation, based on a cooperatively developed and implemented metropolitan-wide strategy, of new and existing transportation facilities eligible for funding under this title and chapter 53 of title 49 through the use of **travel demand reduction and operational management strategies**…	*Strategies that reduce SOV travel and improve existing transportation system efficiency, as produced through the CMP, typically reduce GHG emissions, and could therefore be considered climate change mitigation strategies.*
§134(m)	Additional Requirements for Certain Nonattainment Areas	(1) In general.— Notwithstanding any other provisions of this title or chapter 53 of title 49, for transportation management areas classified as nonattainment for ozone or carbon monoxide pursuant to the Clean Air Act, Federal funds may not be advanced in such area for any highway project that will result in a significant increase in the carrying capacity for single-occupant vehicles unless the project is addressed through a congestion management process.	*Strategies that reduce SOV travel and improve existing transportation system efficiency, as produced through the CMP, typically reduce GHG emissions, and could therefore be considered climate change mitigation strategies.*

Title 23 USC 135 – Statewide Planning (FHWA)			
Section	Aspects	Language	Relation to Climate Change
§135(a)	General Requirements	(1) Findings.--It is in the national interest to encourage and promote the safe and efficient management, operation, and development of surface transportation systems that will serve the mobility needs of people and freight and foster economic growth and development within and through urbanized areas, **while minimizing transportation-related fuel consumption and air pollution**. (3) Contents.--The plans and programs for each State shall provide for the development and integrated management and operation of transportation systems and facilities (including pedestrian walkways and bicycle transportation facilities) that will function as an intermodal transportation system for the State and an integral part of an intermodal transportation system for the United States.	*Provides a link to GHG mitigation through emphasis on minimizing fuel consumption, since GHG emissions from transportation are largely correlated with fuel consumption, and air pollution (1).* *Emphasis is also placed on an integrated transportation system to serve the country (3); States will need to consider the implications of climate change (such as sea level rise) on their infrastructure to ensure effective connectivity is preserved.*
§135(c)	Scope of Planning Process	(1) In general.— Each State shall carry out a statewide transportation planning process that provides for consideration and implementation of projects, strategies, and services that will— (A) support the economic vitality of the United States, the States, nonmetropolitan areas, and metropolitan areas, especially by **enabling global competitiveness, productivity, and efficiency**; (B) increase the **safety** of the transportation system for motorized and nonmotorized users; (C) increase the **security** of the transportation system for motorized and nonmotorized users; (D) increase the **accessibility and mobility** of people and freight; (E) **protect and enhance the environment, promote energy conservation, improve the quality of life, and promote consistency between transportation improvements and State and local planned growth** and economic development patterns; (F) enhance the **integration and connectivity** of the transportation system,	*Requirements for strategies to address safety (B), security (C), accessibility (D), connectivity (F), and preservation (H) will require some States to consider projected impacts of climate change, including sea level rise, on infrastructure. Adaptation strategies will need to be implemented to ensure continued connectivity and accessibility, as well as to promote security of the system, ensure the safety of the system for users, and to support global competitiveness and efficiency (A).* *Requirement of strategies in (E) provides a link to GHG mitigation through emphasis on energy conservation (since GHG emissions from transportation are largely correlated with energy consumption) and consideration of environmental protection.*

			across and between modes throughout the State, for people and freight; (G) promote efficient system management and operation; and (H) emphasize the **preservation** of the existing transportation system.	
§135(d)	Additional Requirements		In carrying out planning under this section, each State shall, at a minimum, consider-- (1) with respect to nonmetropolitan areas, the concerns of local elected officials representing units of general purpose local government; (3) coordination of transportation plans, programs, and planning activities with **related planning activities** being carried out outside of metropolitan planning areas.	*States should consider, as part of the requirement, climate action planning activities going on within their State or region, as well as local government plans or policies that may consider climate change.*
§135(f)	Long-Range Statewide Transportation Plan		2) Consultation with governments.— (D) Consultation, comparison, and consideration.— (i) In general.— The long-range transportation plan **shall be developed, as appropriate, in consultation with State, tribal, and local agencies responsible for land use management, natural resources, environmental protection,** conservation, and historic preservation. (4) Mitigation activities.— (A) In general.— A long-range transportation plan shall **include a discussion of potential environmental mitigation activities** and potential areas to carry out these activities, including activities that may have the greatest potential to restore and maintain the environmental functions affected by the plan. (B) Consultation.— The discussion shall be **developed in consultation with Federal, State, and tribal wildlife, land management, and regulatory agencies**. (7) Existing system.— The statewide transportation plan should include capital, operations and management strategies, investments, procedures, and other measures to **ensure the preservation** and most efficient use of the existing transportation system.	*Direct linkages are possible in consultation with agencies responsible for or involved with climate action planning, including environmental and land use agencies that might also be incorporating climate change considerations – both mitigation and adaptation – into their planning or programs (2 and 4B).* *Mitigations activities are specifically required within the statewide plan (4); climate change mitigation strategies could be directly linked to this requirement.* *Additionally, in considering preservation of the existing system (7), adaptation strategies could be required or encourages.*

23 CFR Part 450 – Planning Assistance and Standards (FHWA)			
Section	Aspects	Language	Relation to Climate Change
§ 450.200	Purpose	The purpose of this subpart is to implement the provisions of 23 U.S.C. 135 and 49 U.S.C. 5304, as amended, which require each State to carry out a continuing, cooperative, and comprehensive statewide multimodal transportation planning process, including the development of a long-range statewide transportation plan and statewide transportation improvement program (STIP), that facilitates the safe and efficient management, operation, and **development of surface transportation systems that will serve the mobility needs of people and freight** (including accessible pedestrian walkways and bicycle transportation facilities) **and that fosters economic growth and development within and between States and urbanized areas, while minimizing transportation-related fuel consumption and air pollution** in all areas of the State	*To ensure a transportation system that will serve the mobility needs of passengers and freight and that fosters economic development between areas, states will need to consider the implications of climate change (such as sea level rise) on their infrastructure to ensure effective connectivity is preserved. Additionally, emphasis on minimization of transportation-related energy consumption and air pollution further reinforces climate change mitigation strategies.*
§ 450.206	Scope of the statewide transportation planning process	(a) Each State shall carry out a continuing, cooperative, and comprehensive statewide transportation planning process that provides for consideration and implementation of projects, strategies, and services that will address the following factors: (1) Support the **economic vitality** of the United States, the States, metropolitan areas, and non-metropolitan areas, especially by enabling **global competitiveness, productivity, and efficiency**; (4) Increase **accessibility and mobility** of people and freight; (5) **Protect and enhance the environment, promote energy conservation,** improve the quality of life, and promote consistency between transportation improvements and State and local planned growth and economic development patterns; (6) Enhance the **integration and connectivity** of the transportation system, across and between modes throughout the State, for people and freight;	*Requirements for strategies to address economic vitality (1), accessibility (4), connectivity (6), and preservation (8) will require some States to consider projected impacts of climate change, including sea level rise, on infrastructure.* *Requirement of management and operations strategies in (7) provides a link to GHG mitigation through emphasis on energy conservation (since GHG emissions from transportation are largely correlated with energy consumption) and consideration of environmental protection.*

			(7) Promote **efficient system management and operation**; and (8) Emphasize the preservation of the existing transportation system.	
§450.208	Coordination of planning process activities		a) In carrying out the statewide transportation planning process, each State shall, at a minimum: (2) Coordinate planning carried out under this subpart with statewide trade and economic development planning activities and related multistate planning efforts; (4) Consider the concerns of local elected and appointed officials with responsibilities for transportation in non-metropolitan areas; (6) Consider related planning activities being conducted outside of metropolitan planning areas and between States; and (7) Coordinate data collection and analyses with MPOs and public transportation operators to support statewide transportation planning and programming priorities and decisions.	*States should consider, as part of the requirement, climate action planning activities going on within their State or region, as well as local government plans or policies that may consider climate change.* *Specifically, the opportunities to link coordinate process to adaptation include ensuring data collection and analysis (7) considers implications of climate change on the transportation system and land use.*
§450.214	Development and content of the long-range statewide transportation plan.		(a) The State shall develop a long-range statewide transportation plan, with a **minimum 20-year forecast**...[that] shall consider and include, as applicable, elements and **connections** between public transportation, non-motorized modes, rail, commercial motor vehicle, waterway, and aviation facilities, particularly with respect to intercity travel. (c) The long-range statewide transportation plan shall reference, summarize, or contain any applicable short-range planning studies; **strategic planning and/or policy studies**; transportation needs studies; management systems reports; emergency relief and disaster preparedness plans; and any statements of policies, goals, and objectives on issues (e.g., transportation, safety, economic development, social and **environmental effects, or energy**) that were relevant to the development of the long-range statewide transportation plan. (i) The long-range statewide transportation plan shall be developed, as appropriate, **in consultation** with State, Tribal, and local agencies	*Potential to link GHG mitigation and adaptation to one of the specific elements listed or to include new element to address.* *Opportunities for linkages to adaptation include the mandate to consider connectivity (a), which requires consideration of the impacts of climate change on transportation infrastructure.* *Reference to "strategic planning and/or policy studies" (c) may include climate change plans or policies.* *Specific reference to environmental effects and energy (c) as well environmental protection and mitigation (l, j) could directly relate to climate change mitigation activities.*

		responsible for land use management, **natural resources, environmental protection**, conservation, and historic preservation. (j) A long-range statewide transportation plan shall include a discussion of **potential environmental mitigation** activities and potential areas to carry out these activities…	
§450.306	Scope of the metropolitan transportation planning process.	(a) The metropolitan transportation planning process shall…address the following factors: (1) Support the **economic vitality of the metropolitan area, especially by enabling global competitiveness, productivity, and efficiency**; (2) Increase the safety of the transportation system for motorized and non-motorized users; (3) Increase the security of the transportation system for motorized and non-motorized users; (4) Increase **accessibility and mobility** of people and freight; (5) Protect and **enhance the environment, promote energy conservation**, improve the quality of life, and promote consistency between transportation improvements and State and local planned growth and economic development patterns; (6) Enhance the **integration and connectivity of the transportation system**, across and between modes, for people and freight; (7) Promote efficient **system management and operation**; and (8) Emphasize the **preservation** of the existing transportation system.	*Requirements for strategies to address safety (2), security (3), accessibility (4), connectivity (6), and preservation (8) will require some States to consider projected impacts of climate change, including sea level rise, on infrastructure. Adaptation strategies will need to be implemented to ensure continued connectivity and accessibility, as well as to promote security of the system, ensure the safety of the system for users, and to support global competitiveness and efficiency (1).* *Requirement to address environmental protect and energy conservation (5) provides a link to mitigation since GHG emissions from transportation are largely correlated with energy consumption and climate change in increasingly recognized as an environmental issue. Additionally, many management and operational strategies (7) are also mitigation strategies.*
§450.320	Congestion management process in transportation	(a) The transportation planning process in a TMA shall address congestion management through a process that provides for safe and effective integrated management and operation of the multimodal transportation system, based on a cooperatively developed and implemented metropolitan-wide strategy, of new and existing transportation facilities eligible for funding under title 23 U.S.C. and title 49 U.S.C. Chapter 53 through the use of travel demand reduction and operational management strategies.	*Integrated management and operational strategies specified in (a) reduce transportation-related emissions by reducing vehicle use or improving traffic flow, and are therefore also climate change mitigation strategies. Congestion management and travel demand strategies typically reduce emissions and therefore link directly to climate change mitigation.*

23 CFR Part 500 – Management and Monitoring Systems (FHWA)			
Section	Aspects	Language	Relation to Climate Change
§500.106	Pavement Management Systems	An effective PMS for Federal-aid highways is a systematic process that provides information for use in implementing cost-effective **pavement reconstruction, rehabilitation, and preventative maintenance** programs and that results in pavements designed to accommodate current and forecasted traffic in a safe, durable, and cost-effective manner. The PMS should be based on the ``AASHTO Guidelines for Pavement Management Systems.''	*In discussing the results of the PMS, there is an opportunity to encourage DOTs to use low-GHG emitting construction materials (such as using fly ash in concrete) as a mitigation strategy.* *There is also an opportunity to require consideration of adaptation strategies to respond to projected climate change impacts in decision-making, in addition to "current and forecasted traffic."*
§500.107	Bridge Management Systems	An effective BMS for bridges on and off Federal-aid highways that should be based on the ``AASHTO Guidelines for Bridge Management Systems'' and that supplies analyses and summaries of data, uses mathematical models to make forecasts and recommendations, and provides the means by which alternative policies and programs may be efficiently considered. An effective BMS should include, as a minimum, formal procedures for: (a) Collecting, processing, and updating **data**; (b) **Predicting deterioration**; (c) Identifying alternative actions;	*Since climate change is expected to cause accelerated aging of infrastructure, particularly bridges, the BMS process could explicitly highlight adaptation concerns in conjunction with (a) data collection, (b) predicting deterioration and (c) identifying alternative actions to encourage consideration of these impacts.*
§500.109	Congestion Management Systems	…The CMS results in serious consideration of implementation of strategies that provide the most efficient and effective use of existing and future transportation facilities. In both metropolitan and non-metropolitan areas, consideration needs to be given to strategies that reduce SOV travel and improve existing transportation system efficiency… (b) In addition…the CMS…shall include: (4) Identification and evaluation of the anticipated performance and **expected benefits** of appropriate traditional and nontraditional congestion management strategies that will contribute to the more efficient use of existing and future transportation systems based on the established performance measures. The following categories of strategies, or combinations of strategies, should be appropriately considered for each area: **Transportation demand management measures, including growth management and congestion pricing; traffic operational improvements; public**	*Strategies that reduce SOV travel and improve existing transportation system efficiency, as produced through the CMP, typically reduce GHG emissions, and could therefore be considered climate change mitigation strategies.* *The evaluation of expected benefits (b4) could also specifically incorporate projected climate change mitigation benefits.*

		transportation improvements; ITS technologies; and, where necessary, additional system capacity.	

49 CFR Part 613 – Planning Assistance and Standards (FTA)			
Section	Aspects	Language	Relation to Climate Change
§613.100	Metropolitan transportation planning and programming	…These plans and programs shall lead to the development of an **integrated, intermodal metropolitan transportation system** that facilitates the efficient, economic movement of people and goods.	*In considering ensuring an integrated transportation system, adaptation strategies could be required or encouraged, since some metropolitan areas will need to consider the implications of climate change (such as sea level rise) on their infrastructure to ensure effective connectivity is preserved. Temperature swings resulting from climate change are also expected to cause accelerated aging on infrastructure.*
§613.200	Statewide transportation planning and programming	…23 CFR part 450, subpart B, requires each State to carry out an intermodal statewide transportation planning process, including the development of a statewide transportation plan and transportation improvement program that **facilitates the efficient, economic movement of people and goods in all areas of the State**…	*In order to facilitate effective future movement, some states areas will need to consider adaptation strategies in light of the implications of climate change (such as sea level rise and accelerated aging) on their infrastructure.*

Title 49 USC 5303 – Metropolitan Planning (FTA)			
Section	Aspects	Language	Relation to Climate Change
§5303(a)	General Requirements	(2) Contents.--The plans and programs developed under paragraph (1) for each metropolitan area shall provide for the development and integrated management and operation of transportation systems and facilities (including pedestrian walkways and bicycle transportation facilities) that will function as an intermodal transportation system for the metropolitan area and as an **integral part of an intermodal transportation system** for the State and the United States.	*To ensure an integrated transportation system to serve the State and the U.S., MPOs will need to consider the implications of climate change (such as sea level rise) on their infrastructure to ensure effective connectivity is preserved. Additionally, emphasis on non-motorized transportation and could also facilitate climate change mitigation strategies.*
§5303(b)	Scope of Planning Process	(1) In general.--The metropolitan transportation planning process for a metropolitan area under this section shall provide for consideration of projects and strategies that will-- (A) support the economic vitality of the metropolitan area, especially by **enabling global competitiveness, productivity, and efficiency**; (B) increase the **safety** of the transportation system for motorized and nonmotorized users; (C) increase the **security** of the transportation system for motorized and nonmotorized users; (D) increase the **accessibility and mobility** of people and for freight; (E) **protect and enhance the environment**, **promote energy conservation**, improve the quality of life, and promote consistency between transportation improvements and State and local planned growth and economic development patterns; (F) enhance the **integration and connectivity** of the transportation system, across and between modes, for people and freight; (G) promote efficient **system management and operation**; and (H) emphasize the **preservation** of the existing transportation system.	*Requirements for strategies to address safety (B), security (C), accessibility (D), connectivity (F), and preservation (H) will require some MPOs to consider projected impacts of climate change on infrastructure. Adaptation strategies will need to be implemented to ensure continued connectivity and accessibility, as well as to promote security of the system, ensure the safety of the system for users, and to support global competitiveness and efficiency (A).* *Additionally, management and operations strategies can often be considered climate change mitigation strategies, if they improve system performance and achieve emissions reductions.* *Requirement of projects and strategies in (E) provides a link to GHG mitigation through emphasis on energy conservation (since GHG emissions from transportation are largely correlated with energy consumption) and consideration of environmental protection.*
§5303(f)	Developing Long-Range Transportation Plans.	…The plan shall be in the form the Secretary considers appropriate and at least shall-- (A) identify transportation facilities (including major	*In considering ensuring an integrated transportation system (A), system preservation (Ci) and mobility and access (Cii), adaptation strategies could be required or encouraged,*

		roadways, mass transportation, and multimodal and intermodal facilities) that should function as an **integrated metropolitan transportation system**, emphasizing transportation facilities that serve important national, regional, and metropolitan transportation functions; (C) identify transportation strategies necessary-- (i) to ensure **preservation**, including requirements for management, operation, modernization, and rehabilitation, of the existing and future transportation system; and (ii) to use existing transportation facilities most efficiently to **relieve congestio**n, to efficiently serve the **mobility** needs of people and goods, and to enhance **access** within the metropolitan planning area;	*since some metropolitan areas will need to consider the implications of climate change (such as sea level rise and temperature fluctuations) on their infrastructure to ensure effective connectivity is preserved.* *Congestion management strategies (Cii) are typically also relevant to climate mitigation, as they can reduce GHG emissions.*

www.ingramcontent.com/pod-product-compliance
Lightning Source LLC
Chambersburg PA
CBHW081901170526
45167CB00007B/3108